RANTHAMBHORE SANCTUARY
DILEMMA OF ECO-DEVELOPMENT

About the Author

Professor H. S. Sharma (*b.* 1945) holds post-graduate and doctoral degrees in geography from the University of Saugar, Madhya Pradesh.

Dr. Sharma is a widely travelled geographer across U.K, Germany, Holland, Canada and Norway and has over seventy papers and a dozen books to his credit. Professor Sharma has supervised nearly two dozen Ph.D.'s on geomorphology, environmental studies, desertification, soil erosion, ecological aspects and sustainable development.

Professor Sharma has successfully conducted several research projects sponsored by U.G.C., Ministry of Environment and Forests, New Delhi, DRDO and NORAD, Norway. Founder Member of many national and international bodies and associations, Professor Sharma has made a significant contribution to the development of geomorphology and environmental studies in India.

Recipient of U.G.C., Career Award in 1979, Dr. Sharma is a Professor of Geography and Director, Indira Gandhi Centre for Human Ecology, Environment and Population studies, University of Rajasthan, Jaipur.

RANTHAMBHORE SANCTUARY
DILEMMA OF ECO-DEVELOPMENT

H. S. SHARMA
Professor of Geography & Director
Indira Gandhi Centre
for Human Ecology, Environmental
and Population Studies (HEEPS)
Jaipur

CONCEPT PUBLISHING COMPANY, NEW DELHI-110059

ISBN 81-7022-817-4

First Published 2000

Published and Printed by
Ashok Kumar Mittal
Concept Publishing Company
A/15-16, Commercial Block, Mohan Garden,
NEW DELHI-110059

Phone : 5649024, 5648039
Fax : 091-11-5648053
E-mail : publishing@conceptpub.com

PREFACE

"Nature provides enough to satisfy every man's need not every man's greed."

— *Mahatma Gandhi*

IN recent times the future of the global system is one of the most important issues of the world. To save the planet and its ecosystem has become the global concern of every responsible citizen of the world. It is said that the economic growth and development will be useless if the environment in which we survive collapses. Human beings, who are the largest and the most developed species of the earth, can be held responsible for the selfish destruction of the environment and the habitat of the other species. Human beings know that their survival depends on the upkeep of nature. Thus, in spite of all human hostilities towards nature, man has always tried to protect nature. The creation of sanctuaries and national parks is an evidence of this. The idea of national parks was to protect the ecosystems of the world for future generations and to conserve the various species of flora and fauna.

The present project aims to study the eco-perceptions of Ranthambhore Sanctuary focussing mainly on the impact of sanctuary on the socio-economic conditions of the local inhabitants and their perception of the idea of the sanctuary.

The present project was assigned to me by the Ministry of Environment and Forests, Government of India, New Delhi under the Man and Biosphere Programme. I owe my sincere sense of gratitude to the Ministry for providing the financial support for the project.

The idea of undertaking the project was initially conceived by Dr. Vasumati Shankaran of IIPA, New Delhi. I express my sincere thanks to her for assisting me in the formulation of this project. I was supported by my research team during the entire duration consisting of Mr. Rahul Saxena, Dr. Narendra Yadav, Mr. Tej Ram Meena and Mr. Davendra Singh.

My sincere thanks are due to all of them. I am thankful to Dr. P.R. Binda for providing me all support in managing this project at every stage and cartographic mapping.

I must thank all my colleagues in the department for helping me in successful completion of this project.

I am grateful to Shri A.C. Sharma, Deputy Secretary to Governor of Rajasthan, Raj Bhawan for managing my field trips to Ranthambhore Sanctuary. I also owe my sincere thanks to Shri S. Choudhary, Field Director, Wildlife Sanctuary, Ranthambhore and his officials for extending fullest support in field work.

I also take an opportunity to express my sincere thanks to Dr. H.S. Mathur, Former Associate Professor of Geography, for Consultancy and to Dr. J.R. Sharma, Director, State Remote Sensing Centre, Jodhpur for supplying IRS data for this project. Dr. R.D. Gupta, Secretary, Institute of Sustainable Development Studies, Jaipur deserves appreciation and thanks for providing TM data which he procured under IMSD project for preparation of geomorphological map of the project area.

At the grassroot level are the farmers and administrative officers of the district who most willingly helped me and endured my frequent encroachment on their time. I thank them all for their cooperation.

H. S. SHARMA

CONTENTS

LIST OF FIGURES

LIST OF TABLES

LIST OF COLOUR PLATES

1

INTRODUCTION

IN recent times the future of the global system is one of the most important issues of the world. To save the planet and its ecosystem has become the global concern of every responsible citizen of the world. It is said that the economic growth and development will be useless if the environment in which we survive collapses. Human beings, who are the largest and the most developed species of the earth, can be held responsible for the selfish destruction of the environment and the habitat of the other species. Human beings know that their survival depends on the upkeep of nature. Thus, in spite of all human hostilities towards nature, man has always tried to protect nature. The creation of sanctuaries and national parks is an evidence of this. The idea of national parks was to protect the ecosystems of the world for future generations and to conserve the various species of flora and fauna.

The present project aims to study the eco-perceptions of Ranthambhore Sanctuary focussing mainly on the impact of sanctuary on the socio-economic conditions of the local inhabitants and their perception of the idea of the sanctuary.

In the last few years what has surfaced as a significant issue is the conflict between parks and people. People's movements at grassroot levels have been emerging not only in India, but all over the world, especially in the developing countries over state-sponsored conservation strategy in the form of national parks and wildlife sanctuaries. These struggles constitute a new genre of social conflict manifesting the assertion of rights by the marginalised populace over natural resources, which are crucial for their sustenance. These collective social actions at micro-level have crucial macro-implications with respect to priorities of the present development paradigm and conservation strategy.

Studies indicate that parks and forest reserves are likely to be one of the major issues of future rural social conflict in many countries. To quote,"It is possible that in certain countries the slightest political change or slackness would be sufficient for the people to invade parks, kill animals or set the forest on fire" (Ghimire, 1991). Hence, the serious implications of the issues both at micro- and macro-levels compel us to think over the problems more deeply and diagnose the crucial dimensions of such 'interventionistic conservation'.

National parks and sanctuaries are prominent components of protected areas and important part of broader conservation strategy. Other types of protected areas include nature reserves, biosphere reserves, etc. The formal goal of protected areas specifically national parks, sanctuaries and reserves has been to preserve plants, animals and micro-organisms so also the total ecosystems. At present, national parks and reserves represent the single most important method of conserving biological diversity worldwide. (Wells and Brandon, 1992).

National Parks and Sanctuaries Conservation in India

While India's biological heritage has been slowly eroded over the millennia, the society has been evolving a number of practices to preserve this diversity. In the hunting-gathering state itself, protection was extended to totemic animals. Early Indian literatures including the Hindu epics, the Buddhist Jatakas, the Panchatantra and the Jain strictures against violence to even lowly life forms, are eloquent proof of the respect accorded to wildlife in ancient Indian culture (Centre for Science and Environment, 1982). In about 242 B.C., Emperor Ashoka's fifth pillar edict gave protection to fish, animals and forests. And before that in the treatise on statecraft called the *Arthashastra* (attributed to Kautilya in 300 B.C.), there is clear reference to the establishment of *abhayaranyas* or forest sanctuaries (Gee, 1962; Rao, 1988; Singh *et al.*, 1990).

India became a votary of conservation of wild species and biological diversity and adopted the concept of national parks and sanctuaries. The first park in India, Corbett National Park, came into being in 1935. Since then there has been enormous increase in the number of such protected areas. Within three decades the number of sanctuaries and national parks multiplied seven and fifteen times respectively.

According to an official source, currently 3.5 per cent of the total land surface of the country is under national parks and sanctuaries and it is proposed to increase the area to 5 per cent (Ministry of Environment, 1992). An expert puts the current figure at 4.6 as under national parks and strictly protected areas (Panwar, 1990). Further calculations show that, at present, about 20 per cent of the total forest area is under national parks and sanctuaries. And this 20 per cent of the area constitutes the country's prime forest land.

While the government takes credit for setting up a large network of protected areas, the local people living in and around these protected areas are relentlessly waging struggles against them, sometimes leading to eruption of violence and overt conflict between the park management and the local communities. The anger is ostensibly directed against the protected area and animals. The kind of conflict is not confined to India only, but common to, by and large, all developing countries (Reti, 1986).

Parks, People and Conservation

Experiences from all over the world, especially the developing countries, show that the establishment and expansion of national parks and protected areas have critical impact at the local level. Social, economic, political and cultural specifications of the country and communities are undermined and a large number of peasants have been expelled from their settlements without being provided with any alternatives. The local people are prohibited from cultivating even their long established, customary fields as soon as the area is gazetted as protected area. On one hand, their traditional customary rights are abrogated and on the other, the local people receive no tangible benefits from the protected area (Reti, 1986; Dang, 1991; Ghimire, 1991; Stephen and Amend, 1992; Jena, 1993).

In India too the story repeats itself. The efforts have not yielded the desired result. While the number of parks and sanctuaries has been increasing over the years so also the area under protected zone, it is coming to notice that tiger population within protected reserves has been dwindling (*The Times of India,* 27 December, 1993) and forest cover declining (Cherail, 1993).

National parks and sanctuaries have substantially contributed to the process of alienation for the poorer sections making their survival far more difficult. This has created another arena of conflict between the local people and the protected area. In many places in our country, the conflict has taken the shape of organised struggle and a good number of such struggles have adopted violent means to assert their traditional rights over the endowments of nature, especially forests and wetlands. Very often conflicts over access and control over natural resources in national parks and sanctuaries become law and order problems and result in physical confrontations between the people and the authorities (Kothari *et al.*, 1989). A frenzied mob set fire to about 10,000 acres of Nagarhole National Park, a well-protected Park in Karnataka, in retaliation against the alleged murder of a local person. Elsewhere, in Kerala's Silent Valley and Madhya Pradesh's Kanha Tiger Reserve, similar cases of arson have threatened to undo years of protection (Bagla, 1992; Jena, 1993). Gujjars in Rajaji National Park (Uttar Pradesh), Kolis in Bhimashankar Sanctuary (Maharashtra), Maldharis in Gir National Park (Gujarat), and elsewhere people are questioning the very need and rationale of national parks and sanctuaries.

The major reasons advanced by the officials for this conflict is the efforts by the forest officials to stop illicit felling of trees, poaching, illegal grazing, encroachments and other forest offences (Kothari *et al.*, 1989). Hence, one has to look beyond the legal definition of forest offences and explore the causes and analyse them, within the context of existential situation of local communities.

Parks *Vs.* People's Rights: The Question of Survival

There are two perspectives on forests with regard to its conservation and use. One, which can be called macro perspective, treats forests as the repository of 'precious' biological diversity and wilderness essential for advancement of knowledge and solution to various global problems such as greenhouse effect, climatic change, etc. The second perspective which can be called micro perspective sees forest as crucial endowments to nature directly and indirectly providing sustenance not only to wild species, but also to a large number of people whose social, economic, cultural, political and spiritual life revolves around it.

In India, forest assumes importance in the context of widespread rural poverty, which needs to be seen in context of biomass scarcity prevailing in the country — the shortage of food, fodder, firewood, artisan raw materials, timber, etc. A surplus of these materials will not turn India into a wealthy nation in the modern sense, but it will definitely help to reduce the existing rigours of poverty — the extraordinary work burden that the poor face today. A large section of people in India both tribal and non-tribal, including nomads, depend on forests and their natural surrounding for survival. Livelihood of a large number of artisans is inextricably linked with forests.

Unlike the Western countries, in India one finds a large number of people living with these forests and sometimes in dense forests. For them forest is the only source of livelihood. A study reports that 10 sanctuaries have a population density higher than the national figure and some national parks have very high density, for example, Bansda in Gujarat has 0.56 persons per ha (Kothari *et al.,* 1989). At the same time, a large number of people are found to be inhabiting the areas adjacent to parks and sanctuaries. Kothari *et al.* (1989) have reported more than 80 per cent of their sample sanctuaries and national parks as having human habitation within 10 km radius of the boundary. The population density in the periphery of Ramnabagan Sanctuary (West Bengal), and Karnala Sanctuary (Maharashtra) is very high as 26,675 and 2,232 persons per ha respectively.

Earmarking an area crucial for survival of a sizable number of marginalised people exclusively for rearing and restoration of wild animals and their habitat denying access to them has been a major source of conflict between national parks, sanctuaries and the local communities.

Threat to Right to Life and Livelihood

There is another dimension of the problem which is not a direct consequence of the law but emanates from the protected area. The Status Report on Management of National Parks and Sanctuaries in India (Kothari *et al.,* 1989) reports, "One of the major problems inherent in the management of India's national park and sanctuaries is that very often some of the wild animals sought to be protected adversely affected the interest of the local people. Wild animals causing injury to or death of human beings and cattle, and extensive damage to the crops, are common examples of this. Such incidents occur frequently in and around a number of parks and sanctuaries."

Though, some of the parks and sanctuaries have compensation scheme for such damage, yet the repercussions of such damage for the poor person are very severe. Preponderant instances are found that in some areas people have given up cultivating wheat and other Rabi crops because of regular damage caused by the wild animals (Jena, 1993;

PRIA, 1993). As a result, already poverty-stricken families lose their food crops and starve. As far as compensation is concerned, it involves legal intricacies and bribe to the officials, hence, in most of the cases, even if the person is aware about the compensation scheme, he simply forgoes it. It is clear from the above delineation that the contemporary state-sponsored conservation is based on elitist perceptions and ignores the micro-realities. 'Interests' and traditional rights of the local people living in and around national parks and sanctuaries have become the biggest casualty.

While the official ideology envisages that the goal of conservation is sustainable development, but in reality it has unleashed the process of deprivation for a large number of people. It is further clear that sustainable development through conservation is devoid of an essential component, i.e. sustainable use. The very concept of sustainable development becomes meaningless and irrelevant without adequate emphasis on people's needs, rights, knowledge and participation. The result is before us. Parks and sanctuaries cannot exist as insulated islands, ignoring the local people and their survival needs. Biological diversity cannot thrive with cultural diversity. Hence what is required at the moment is review and revision of the approach to and strategy of conservation both at international and national levels.

Need of the Project

The objective of the project was to study the impact assessment of Ranthambhore Sanctuary. The need to take up detailed study arose on account of some very strong ecological reasons of far-reaching consequences. The reasons can be enumerated as follows :

1. A fast-growing human and livestock population (2.00 lakhs and nearly 1.00 lakh respectively) in 97 villages that surrounds the Ranthambhore Tiger Reserve including the two townships of Sawai Madhopur and Khandar has been continuously causing a mounting pressure on the limited resources of agricultural land, forests and wildlife.

2. Increasing developmental and urban activities have been fast impairing the delicate ecological network and the local environment to the level of impossible halting.

3. The human and livestock population depend upon the nearby adjoining forest areas for fuel, fodder, grazing and small timber. Increasing exploitation of forest resources has severely denuded the

sanctuary and resulted in conflict between the local people and forest officers. As a result the ecological components in the Ranthambhore Sanctuary appear to be deteriorating in all respects. This needs to be studied in detail.

4. Since government's policy towards tourism is to earn more and more revenue, there has been fast increase in religio-tourist influx in the sanctuary. This seasonal tourism has given rise to a totally new system called "Tourism culture". This tourist culture has a high physico-chemical mode of life and behaviour, thereby the life-sustaining raw materials are not by and large environmental friendly, but mostly biologically non-degradable synthetic, metallic and artificial. This has altered the ambient environmental status and resource management requires a detailed study.

5. After the declaration of Ranthambhore sanctuary as a National Park in November 1980, 12 villages from the core area of the sanctuary have been shifted and resettled in two villages far apart from each other namely Kailashpuri and Gopalpura. This uprooting has adversely affected the life-style of traditional pasturalists. Four villages are still in the core area and are in the process of evacuation. The extent of their social impacts requires to be studied.

6. The unique geomorphological set-up of Ranthambhore Sanctuary always tempted the principal investigator to take up a study in this zone on the basis of satellite data and detailed field investigations in order to assess the water potentialities of this fractured zone.

It was in this background of thinking and existing scenario of the natural forest resources that this project was conceived and undertaken.

Objectives of the Study

In view of the above background the study was designed to meet the following objectives :

1. The Ranthambhore Sanctuary is an ecological island in the vast degraded surroundings, with biotic pressure for firewood, timber and grazing. The local people are basically pasturalists and agriculturists and dependent on the forests of the sanctuary for their basic needs. Therefore the major objective is to assess the ecological status of the Ranthambhore Sanctuary and to study the socio-economic conditions of the local people.

2. For any eco-development intervention and policy framework a better grasp and understanding of people's perception, responses, priorities and their participation is therefore undoubtedly called for.

3. Tourists activities are concentrated within the core area of the sanctuary leading to high concentration of tourists. There is additional disturbance from pilgrims to the Ganesh temple in the Sanctuary.

4. To study the temporal negative and positive changes in the status of biomass on the basis of satellite data.

Methodology

The methodology of the present study was designed with the basic objective of studying the impact assessment of the Ranthambhore Sanctuary. The study is based on field investigations, descriptive and quantitative analysis of the basic data and information collected from various government departments, reports and documents. Primary data were collected having field work of 7 villages in and around the National Park. A period of 3 to 4 months was devoted to field work in different seasons to have a deeper insight of man-nature relationship at grassroot level and to analyse it both qualitatively and quantitatively. Nearly 97 villages surround the Ranthambhore Tiger Reserve, 54 villages are located in and around the Ranthambhore Sanctuary. Therefore 54 villages including 5 within the core area and 2 newly resettled villages have been selected for collection of primary and secondary data. Primary data of each household of 5 villages (Indala, Mordoongari, Katholi, Padra and Bodal) were collected. Twenty percent sampling from 2 newly resettled villages has been done in order to assess their perception and problems of rehabilitation. The structured schedules were designed for collection of primary data (appended), interviews of villagers, forest guards and visitors. Two IRS LISS 11 scenes i.e. 29/42 A2 and B2 for the 1992, 1994 and 1996 with a 3 metre resolution have also been used. The data were acquired from the State Remote Sensing Centre, Jodhpur. Geomorphological map has also been prepared using satellite data and field checks. Slope map has been prepared with the help of Wentworth's method of average slope analysis.

Review of Literature

A significant study on biodiversity conservation through eco-development and indicative plan for various Tiger Reserves of India including Ranthambhore Tiger Reserve, Rajasthan sponsored by UNEP, has been conducted by ILPA, New Delhi (1994). Grazing and cutting pressure of Ranthambhore National Park has also been done by Berk Muller, Mukherjee and Mishra in 1998. Some reports of forest department are also available which incorporate the status of forests of Ranthambhore National Park. Valmiki Thaper of Ranthambhore Foundation has also carried out extensive review of Ranthambhore National Park. Apart from the above cited references a plethora of related literature has been studied in order to have a current trend of research relating to the theme of the project. A detailed bibliography has been given at the end of this report.

Organisation of Study

The study has been divided into five chapters. In introduction, review of literature, objectives and methodology have been discussed. In second and third chapters, physical, social and economic environments have been discussed. Fourth chapter is the main theme of the project analysing the impact assessment of the Sanctuary. In the last chapter conclusions and recommendations have been included.

2

PHYSICAL ENVIRONMENT

Location

SITUATED amidst the high ranges of the Aravallis in between the confluence of the Chambal and Banas rivers (Fig. 2.1), the Ranthambhore National Park occupies a pride place in Rajasthan for conservation of wildlife for more than a century. Today, it is one of the major centres of Project Tiger in Rajasthan where besides the dwindling species of the Indian tigers, other wild animals like Sambhar, Chinkara, Neelgai, antelopes, wild boars, civets and many others find a natural home. Ranthambhore, besides having a national park of global importance, has an ancient fort, the silent spectator of many a heroic battle between the Rajputs and the Muslims, and a bastion of ever memorable tales of chivalry and bravery.

The Ranthambhore Park, located to the east of the Sawai Madhopur town, about 15 km away, and approachable by an all-weather motorable road is girdled by Aravalli ranges with precipitous slopes, flat-topped crests and cascading streamlets, providing an ideal location for wildlife conservation. The Aravallis vary from 350 m to 450 m high above sea level and are well-clad with forests and natural vegetation of grasses, shrubs and endemic trees — a true ecotone between the semi-arid north and the sub-humid south. The Banas in the north and the Chambal in the south collect the waters of these hills, with most of the streams being seasonal but carrying huge flows of water in times of heavy downpours.

Prior to the declaration of this wildlife reserve as a National Park, the area was a game reserve of the former feudal kings and Maharajas of

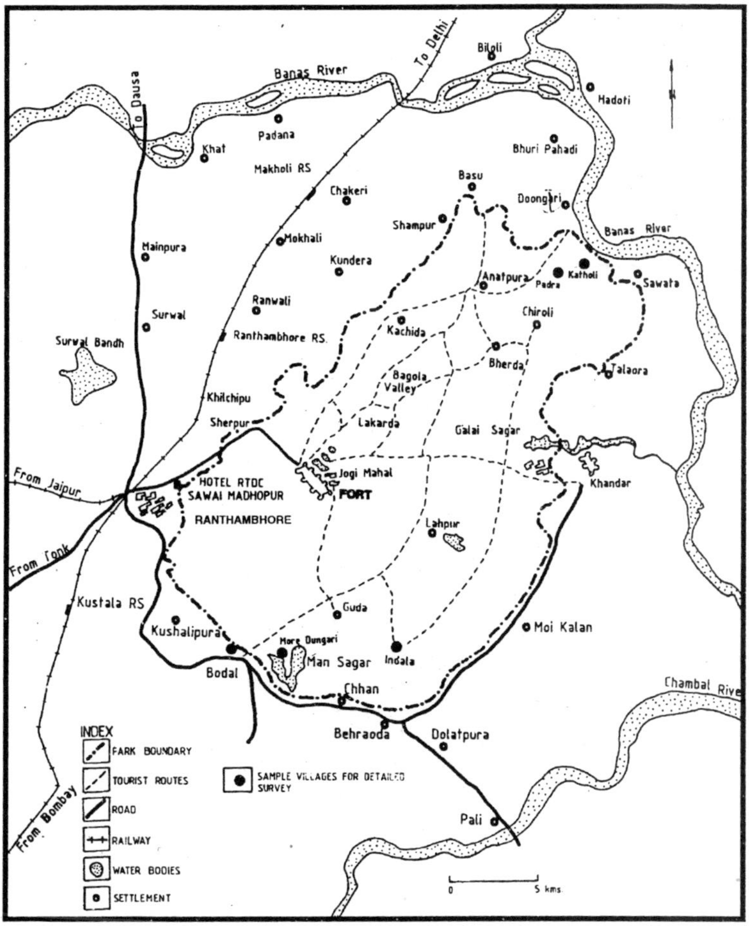

FIG. 2.1 : *Ranthambhore Sanctuary : Location Map*

the Jaipur State. Till that time about 17 villages located within the Park had all their human activities inside the Park, including agriculture, livestock-rearing, irrigation and migration through narrow bridle paths. But after the declaration of the area as a National Park, all the villages, except four, have been evacuated. The four villages that still exist inside the park are Padra and Katholi in the north-east and Indala and Mordoongri in the south near Man Sagar lake. Perhaps, the fertile land of these places coupled with the easily available means of irrigation in the close proximity have deterred the inhabitants of these four villages to migrate out of this wildlife reserve.

Sawai Madhopur town is well-connected by railway network and metalled roads. The town is connected with Jaipur, Delhi and Kota by broad-gauge railway line, providing easy accessibility to Jaipur — the capital of Rajasthan State, Delhi — the capital of India and Mumbai the main commercial centre of the country. Thus owing to easy accessibility, Sawai Madhopur today is a big centre of tourists, nature lovers and wildlife conservationists besides its historical and mythological importance.

The entire area of the Ranthambhore Sanctuary (392 km^2), having the remnants of old abandoned villages, is criss-crossed by fair weather jeepable roads and bridle paths.

Inside the core area are a few natural ponds and depressions notable among them being Man Sagar, Lahpur pond, Milak Talao near Jogi Mahal and Galai Sagar near Khandar.

Outside the core area of the Park, the fertile alluvial land of the Banas in the north is dotted by a large number of villages, mostly along the railroad to Delhi. The RTDC Hotel outside the precincts of the Park and the Jogi Mahal Hotel inside the Park are notable man-made structures, today having a flourishing tourist industry.

Inside the core area, remnants of the old villages like Anatpura, Chiroli, Kachida and Bherda in the north, Lakarda and Lahpur in the middle and Guda in the South exist. The rudiments of the old buildings and the 'Sikargahs' of the former Maharajas of erstwhile Jaipur State still exist.

The surrounding villages along the periphery of the core area and the buffer zone are many in number. To the north of the Ranthambhore Railway Station exist the villages of Ranwali, Kundera, Mokhali, Chakeri and Shampur. Still north, on the right banks of Banas river lie villages like Khat, Padana, Bhuri Pahadi, Basu and Doongri, all having existence due to the fertile cultivable alluvial land of the Banas. Two villages viz. Surwal and Mainpura lie along the Dausa road which is motorable throughout the year.

The existence of a few villages on the eastern margins of the sanctuary viz. Sawata, Talaora, Moi Khan and Khandar is again due to the existence of fertile alluvium of Banas and Chambal rivers, which make a confluence in this part and the river inundations for most of the year spread layers of newer alluvium year after year.

In the south, the villages exist either in the immediate periphery of the zone like Chhan, Behraoda, Bodal and Kushalipura or along the tarred road like Dolatpura and Pali. Thus the surrounding villages of the sanctuary owe their existence to the locational advantages like the existence of metalled roads, railroads and the rich fertile alluvial soil. It is likely therefore that these villages are richly embedded to their agrarian economy and rural trade.

Administrative Units

The Ranthambhore Sanctuary falls in two tehsils, the Khandar in the north and east and Sawai Madhopur in the west and south, the former having a larger area (Fig. 2.2).

The villages and towns surrounding the sanctuary lie in both the tehsils, administered by a common district headquarters — the Sawai Madhopur Town, which lies to the immediate west of this sanctuary, and has a broad-gauge railway station with connections to Kota and Mumbai in the south, Jaipur in the west and Delhi, Bharatpur in the north. Sawai Madhopur — an old town of the former Jaipur feudal state — is surrounded by low hills of the Aravallis called the 'Dangs' and had the special reputation of dense vegetation and grasslands which promoted the cottage industries of wooden toys, Khus scent, Khus fans and various handicrafts. The Chambal and the Banas rivers circumventing the entire area provided rich alluvial soil, easy availability of ground water at shallow depths and storage of surface water in natural tanks and nullahs.

It is easy therefore to understand that the entire area in and around the sanctuary is thickly inhabited by people — most of whom are 'Meenas' a scheduled tribe of Rajasthan who unlike other tribes of Rajasthan State, are well-settled and practise indigenous agriculture and animal husbandry since the earliest times.

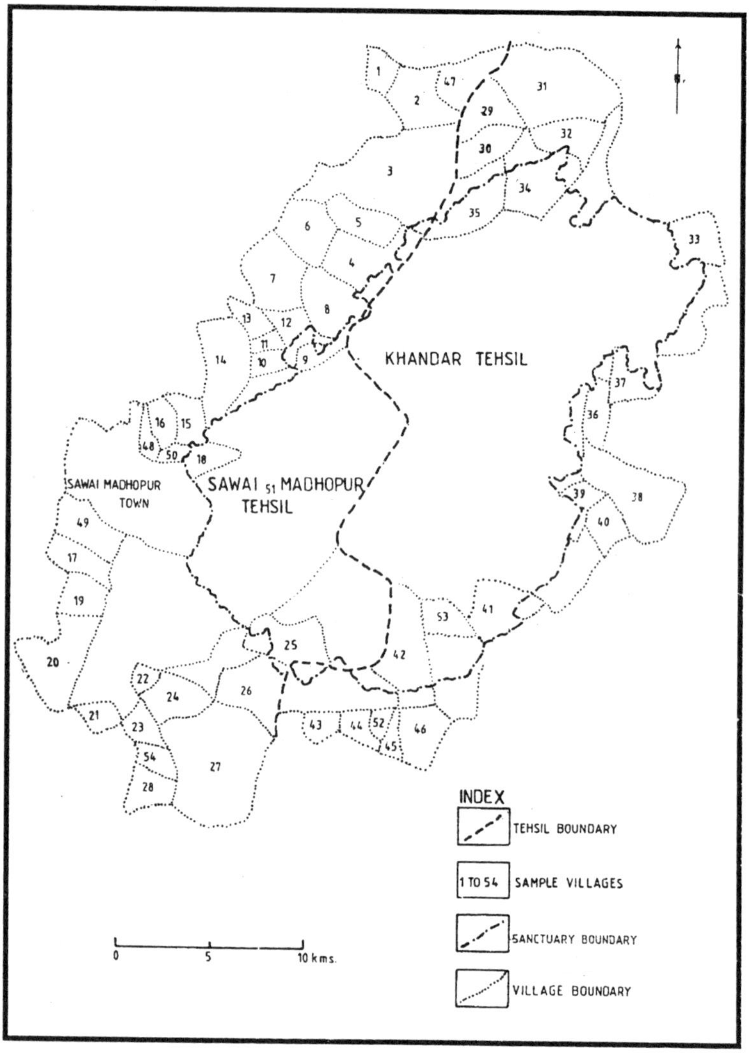

FIG. 2.2 : *Ranthambhore Sanctuary : Administrative Units*

The Ranthambhore Sanctuary is today surrounded by 54 villages (their names, population and occupations have been given in Table 2.1). Two villages viz. Kailashpuri and Gopalpura have now been added, which have been newly built to rehabilitate the displaced persons of those villages which were formerly inside the sanctuary.

TABLE 2.1 : **Ranthambhore Sanctuary and Environs**

Sample Villages

S. No.	Villages	S. No.	Villages
1.	Olwara (S. M.)	28.	Dumoda
2.	Endwa	29.	Baso Kalan (Khandar)
3.	Shyampura	30.	Baso Khurd
4.	Parli	31.	Bhoori Pahari
5.	Bhadlao	32.	Dungri
6.	Kundera	33.	Sawtan
7.	Rawal	34.	Katholi Dhanapala
8.	Uliyana	35.	Padra Nizamat
9.	Khandoj	36.	Goth Bihari
10.	Sherpur	37.	Sanbar
11.	Kutalpura Maliyan	38.	Khandar
12.	Khawa	39.	Rawrah
13.	Jamool Khera	40.	Jaisinghpura
14.	Khilchipur	41.	Mei Kalan
15.	Ramsinghpura	42.	Indala
16.	Khatipura	43.	Jetpur
17.	Borif	44.	Chhan
18.	Madhosinghpura	45.	Allapur
19.	Nimli Khurd	46.	Bahrawada Khurd
20.	Nimli Kalan	47.	Niwandi (S.M.)
21.	Kalibhat	48.	Bhundeda
22.	Kushalipura	49.	Jeenapur
23.	Hajjam Kheri	50.	Himmatpur
24.	Halonda	51.	Qila Ranthambhore
25.	Mordoongri	52.	Sakhwas
26.	Bodal	53.	Gopalpura
27.	Bheronpura	54.	Kailashpuri (S.M.)

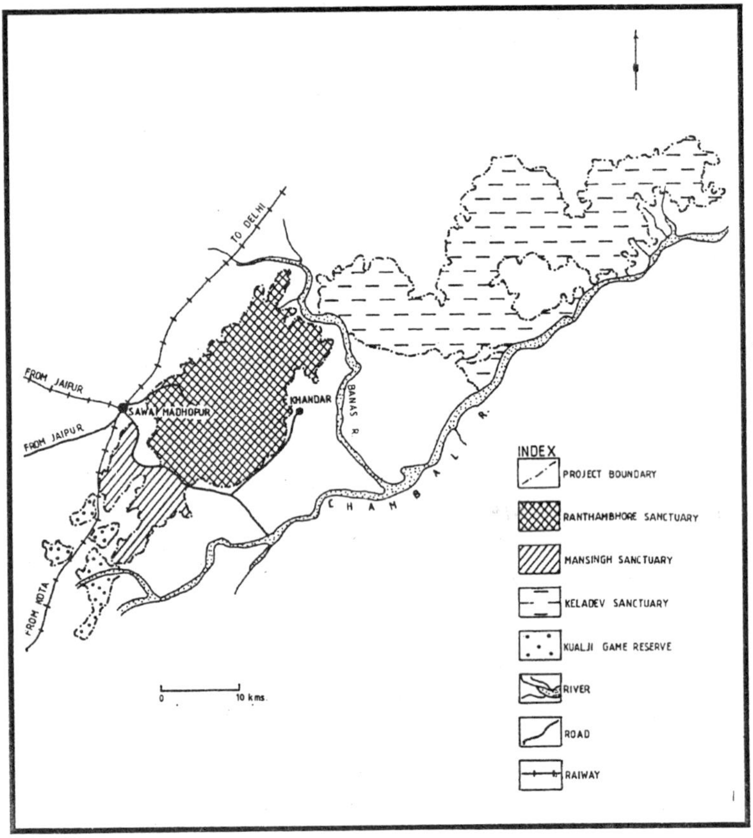

FIG 2.3 : *Ranthambhore and Environs : Parks and Sanctuaries*

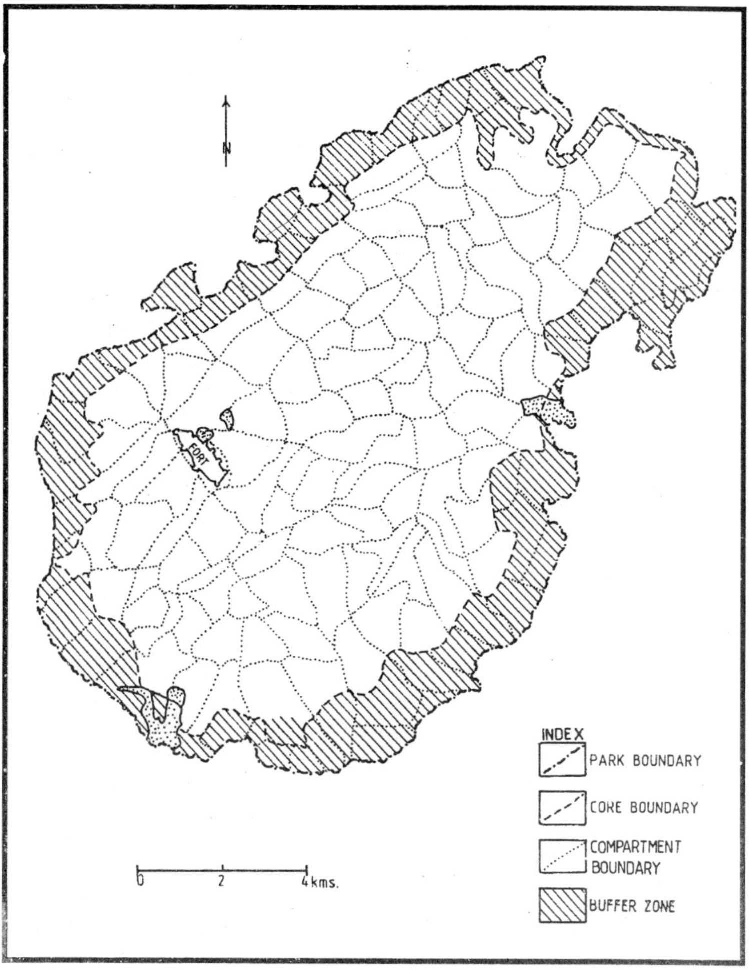

FIG 2.4 : *Ranthambhore Sanctuary — Core and Buffer Zones*

Park and Sanctuaries

The Ranthambhore National Park, towards the end of the last decay, was joined to other game reserves, sanctuaries and reserved forests in order to provide a buffer zone to the main park and thus reduce the grazing pressure of the wild and domestic animals and also to provide a contiguous corridor to wildlife for their better genetic breeding. Thus the reserved forests which existed on the left banks of the Chambal river, to the north and south-east of the main core area were given the status of sanctuaries and game reserves. This contiguous zone now spreads to more than 100 km area from north-east to south-west under different names.

The most prominent of these extended areas is the Kailadevi Sanctuary spreading to the north of Banas river and having a rich growth of trees and grasslands for the wildlife. To the south-west lies the Mansingh Sanctuary, separated from the Ranthambhore core by the Jaipur road. Still further to the south along the railroad to Kota are the Kualji game reserves, spreading out in the four isolated patches (Fig. 2.3).

Formation of these adjoining sanctuaries and game reserves has indeed added to the ecology of the area, giving opportunities to the wildlife to intermingle instead of being caged amidst the plethora of hills. It will also have an added impact on the tiger population, which needs large areas to reign upon and feed themselves. The grazing pressure of domestic animals of the nearby villages will also be lessened when these sanctuaries and closed areas are periodically opened for grazing and cutting of grasses, specially after the rainy season.

Core and Buffer Zones

The core area of the Sanctuary is 274.50 square kilometres which is surrounded on all sides by a buffer zone (118.00 square kilometres). Thus the total area under the Ranthambhore Sanctuary is 392 square kilometres (Fig. 2.4).

The width of the buffer zone is obviously determine by the surrounding ranges which girdle the entire sanctuary from all sides and make it impregnable for the wild animals to escape out. Still, at times, the tigers come out of the sanctuary and move around the close vicinities of the Sawai Madhopur Town. For the human and cattle population to move inside, there are a few exits, which were formerly used by the native population of the villages existing inside this area but have now been 'evacuated'. These entrances are often transgressed by villages to feed their cattle population on rich grasses that abound mostly in the rainy season. The impact of these frequent encroachments, which often lead to feuds between the Forest personnel and the villages, has been discussed at length in later chapters.

Blocks and Compartments

For the sake of maintenance and proper care, the entire sanctuary has been subdivided into 5 blocks (Fig. 2.5) :

1. Sawai Madhopur — A Block on the west and north-west.
2. Sawai Madhopur — B Block on the south.
3. Khandar — A Block on the north.
4. Khandar — B Block in the middle.
5. Khandar — C Block in the south-east.

Each of the above 5 blocks has been further subdivided in several compartments. Neither the blocks nor the compartments are of same size, rather they have been delineated on the basis of natural barriers like mountain spurs, streamlets, plain lands, valleys etc.

The number of compartments in each block is as below :

1.	Sawai Madhopur	A Block	47 compartments.
2.	Sawai Madhopur	B Block	20 compartments.
3.	Khandar	A Block	39 compartments.
4.	Khandar	B Block	22 compartments.
5.	Khandar	C Block	36 compartments.
	Total		164 compartments.

Each forest guard, patrolling officer and rangers are allotted some compartments for maintenance, vigilance and counting of animals and thus are responsible for both the vegetative and the animal growth. Still wood-cutting, lopping of trees and shrubs, cutting of grasses are an everyday common feature and these lapses may either be with the connivance of forest guards or in the night when the whole forest is in the

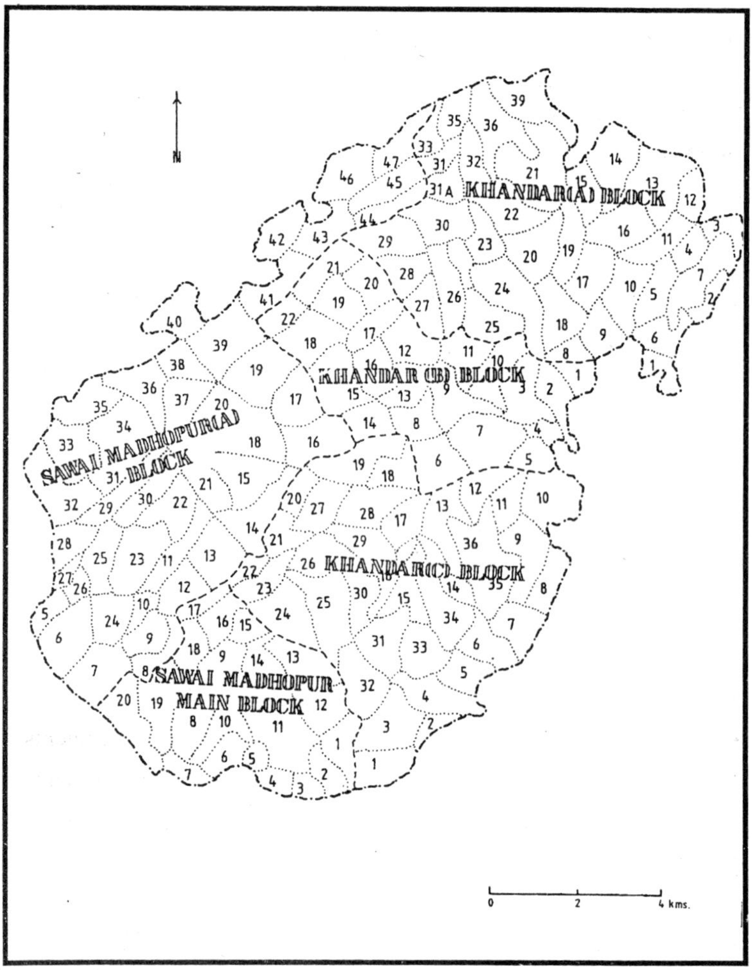

FIG 2.5 : *Ranthambhore Sanctuary — Blocks and Compartments*

shadow of darkness. But still some vulnerable points can be put at constant vigil and these thefts can be arrested.

Climate

Climate is subtropical with distinct cold, hot and rainy seasons. Summer months are from April to June, rainy months are from July to September and winter months are from October to March. Temperature varies from a maximum of about 47°C in May-June to minimum of 4°C in January.

Humidity

Relative humidity is generally over 60% during the south-west monsoon. During the rest of the period the air is dry. The relative humidity is extremely low in May averaging only 37%. But it starts increasing by the middle of June due to effect of pre-monsoon showers, when it may be around 66%. During the rainy season it exceeds 75% but in winters again it ranges from 45% to 60%. The humidity varies from place to place. In the well-crowned valleys it is always higher where the animals come for a siesta during the day and it is always lower on the crests of the hills which are always exposed to sun and winds.

Rainfall

The average annual rainfall is about 68.73 cm. But it is highly variable. While in 1986, 1987, 1989 and 1990, the annual rainfall was less than the average, from 1991, it shows the positive trend. In 1996, the annual rainfall was 90.20 cm (Table 2.2). 92% of precipitation occurs during the period June to September. Droughts are also not uncommon.

The rivers and water holes start drying up soon after the rains and become boggy by the end of winters leaving little hope for respite to the wild animals. The grasses also thin out in the low rainfall months and by the beginning of March, the Park appears to be barren and devoid of forage to the animals.

Winds

Winds are light to moderate. Strong winds in summers and early south-west monsoon season blow mainly from westerly and south-westerly direction. Thunderstorms in May to September and duststorms during summer are common. Cold waves from north-north-west direction blow during winter.

TABLE 2.2 : **Annual Rainfall**

Average Rainfall = 68.72 cm

Year	Annual Rainfall (cm)	Variation from Average Rainfall (cm)
1986	67.80	-0.92
1987	43.00	-25.72
1988	71.60	+2.88
1989	35.40	-33.32
1990	40.40	-28.32
1991	77.30	+8.58
1992	80.70	+11.98
1993	71.30	+2.58
1994	113.00	+44.28
1995	108.60	+39.88
1996	90.20	+21.84

Source : Meteorological Department, Govt. of Rajasthan, Jaipur.

The rainfall, temperature and humidity invariably affect the distribution of wild animals and birds. These climatic factors influence the seasonal migration of these animals. However, in the pinch periods of May and June when the sources of water are limited to only a few spots where perennial supply is available, the animals move in herds in search of water and the hydrology of the area is affected.

Hydrology

Availability of water is the most potent factor in the distribution of animals and plants. Water is the main sustenance of life here as everywhere. During the monsoon season a large number of waterfalls cascade down the cliffs providing the most spectacular and exquisite beauty to the area. As there are a large number of vertical cliffs in this Park, water flows down the waterfalls very soon even during the small span of rainfall for a day or two. The Park has six reservoirs of water that have water throughout the year. Galai Sagar and Mansarovar reservoirs have been artificially created by damming the rivers and they supply water for irrigation to nearby villages. Due to human activities at these points the wild animals seldom visit these sites. It is only during the peak summer season, when other water holes get dry, that the animals come to these sites.

The other four tanks are natural in which the rain water collects into the natural depression. Padam Talao is a vast shallow lake on the banks of which the famous Jogi Mahal, the tourist hotel, stands. It has water round the year. Herds of wild animals visit this lake every morning and evening giving a picturesque view to the pleasure of the visiting tourists. This lake had thick vegetation of grasses and trees around it which provide shelter and food to the animals. Raj Bagh is another big lake having water throughout the year. This lake is quite close to Padam Talao but is less extensive in area. Milak Talao is another large depression to hold rain water but sometimes when the summers are blistering and the arrival of monsoon is delayed, then this lake gets dry. Another small lake is Lahpur lake formed by damming the Lahpur river. The embankments are not high and excess of rain water spills over the surrounding areas causing marshes. The lake dries up in February and provides little respite to the wild animals.

Nonetheless, about a hundred water holes exist in the Park that provide water to animals in the pinch periods. Some of them are scattered pools, once constructed by the rulers of the fort but are now in ruined condition. They form a part of the ecosystem of the Park and serve as valuable water reserves. Other water holes are either depressions in the plateau region or the cesspools of water in the narrow valley (Fig. 2.6). A list of these water holes is given in Table 2.3.

In 1989 and 1990 the rainfall was 35.40 cm and 40.40 cm respectively, which was less than the average rainfall of 68.72 cm. Therefore the acute shortage of water caused crisis of unimaginable magnitude. Most of the water holes and river valleys had dried up in the parching heat of 1989-90 and the herds of animals moved together in search of water from pillar to post but to no avail. Even the major reservoirs, Milak-Talao and Lahpur lake, reputed to contain water all the year round had parched up into sun-baked tiles of mud. The management at that time deepened the existing wells and pumped out water from them to fill in the artificially built, small tanks of water to make possible the survival of animals.

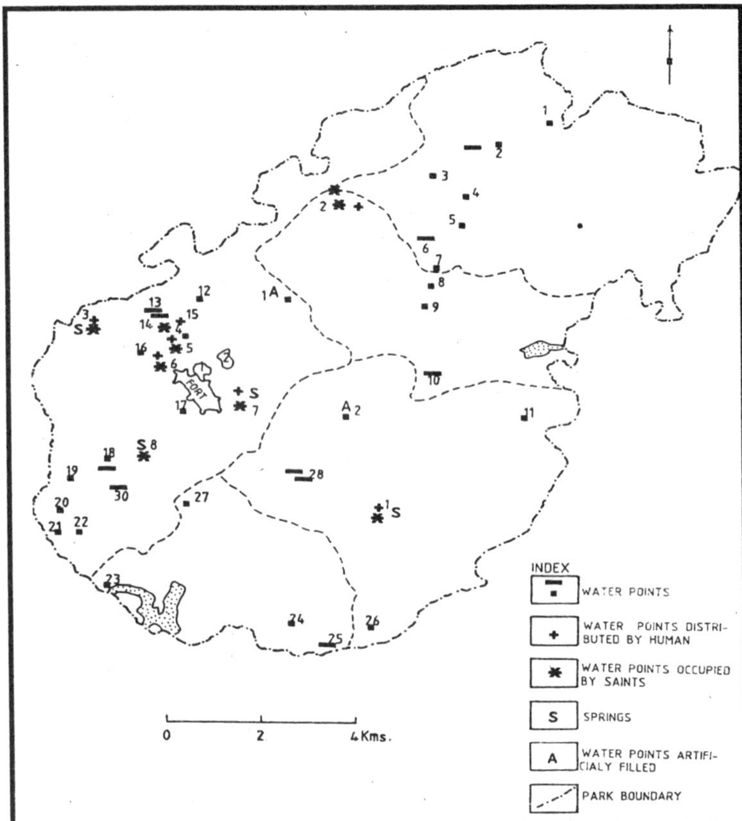

FIG. 2. 6 : *Ranthambhore Sanctuary — Main Water Holes in Summers*

TABLE 2.3 : **Water Points Available in Summers**

1.	Ranideh	16.	Chulideh
2.	Dundarma Ka Dara	17.	Khenshya Kund
3.	Piplideh	18.	Kalapani
4.	Ananthpura	19.	Sultanpur Ki Kui
5.	Baghdeh	20.	Kamdama
6.	Bhakula	21.	Patwabawri
7.	Gulardeh	22.	Khirai kao
8.	Beheda	23.	Niladeh
9.	Bhawardeh	24.	Indala Ki Bavri
10.	Sagariki Khal	25.	Binda Krakhal
11.	Bhanwar Khho	26.	Gopal Kund
12.	Nurkya	27.	Gandlyadeh
13.	Mata Khora	28.	Odikha
14.	Dhobighat	29.	Pila Pani
15.	Singhdwar	30.	Hathipura

Water Points disturbed by Man

1. Khataula
2. Kachida
3. Amreshwar Mahadev
4. Miserdara
5. Ada Balaji
6. Markund
7. Kamaldhar
8. Saleshwar
9. Mansarowar

Water Points occupied by 'Babas'

1. Amreshwar Mahadev
2. Kamaldhar
3. Saleshwar
4. Khataula

Water Points artificially filled

1. Lakarda
2. Lahpur Chauki
3. Padam Talao
4. Raj Bagh Talao

After 1990, the annual rainfall was above the average rainfall, due to which such an acute problem had not been noticed. In 1996, the annual rainfall was 90.20 cm which was above the average rainfall by 21.48 cm.

A survey of the existing water holes shows that most of them are small, of the size of 10' x 10' or even less than this. Water holes of perennial flowing water are few, such as the Sitamata spring, Pir Khera, Kandull, Mishar Darrah, Tapkan, the Banas and the Chambal, all of which are in Sawai Madhopur A and B blocks. Water holes of larger size are Kushlipura, Devpura, Dualada Ka Jharna, Khemsya Kund, Hati Patta, Kirni Khora and Bhata Ki Pass Nala. The smaller water holes disappear gradually with the advancement of summers and the rise in temperatures. Besides causing distress to the animals, the diminishing water supplies affect the density of vegetal growth and the fishes living in the water holes. When the pumps of the wells are recommissioned to fill in the artificial pools of water, herds of Cheetals, Nilgais and Sambhars come down to these sites, line up along the convey drains and wait for their turn to get their share of water, presenting a spectacular sight.

Observations indicate that the majority of wild animals concentrate around Padam Talao, Rajbagh lake, Lahpur lake and Nalghati water holes which have abundance of water throughout the year. Nearly 18 water holes remain disturbed due to the visiting tourists, pilgrims visiting the mosques and temples and the villagers trespassing these areas. These human activities are enough to show off any wild animal.

The Chinkaras prefer to inhabit the peripheral areas and high slopes of the hills and in these areas water holes are scarce. They descend down the hills to fulfil their daily requirements of water. The Kachida valley is the home of shy animals like panther and sloth bears and here too, water holes are scarce. There is a need to augment the water supplies in this valley to support the wildlife system. While it is true that no artificial supply of water can be brought here and the wilderness of the area has to be preserved to form its own ecosystem and natural equilibrium, it is significant that the available water should be harnessed to the maximum and the life be rejuvenated in lean summer months.

Geology and Lineaments

Geologically, the area is distinctly divided into two major units. The eastern side of the study area is occupied by the Vindhyans (mainly shales) which are mostly horizontal in nature without much tectonic disturbances. The western hill ranges of the centre of Ranthambhore Sanctuary belong to the Aravalli group of rocks which are folded and fractured (Fig. 2.7).

Pre-Aravalli rocks comprise quartzites, mica schists, gneisses and migmatities. The rocks of Vindhyan super group comprise sandstones, shale, limestone and breccia of widely variable composition and character occupying most part of this reserve. They are separated from the older rock units namely Pre-Aravallis and Delhis by a reverse fault trending north-east, south-west called the 'Great Boundary Fault'.

An alternate succession of sandstone and shale with intervening carbonate phase and their associated primary structures of the Vindhyans reveal a shallow marine environment deposition with signs of mid tectonism. Similarly the alternate sequence of quartzite and phyllites within Delhis essentially reflects stable tectonic framework of sedimentation with alternate period.

The Aravalli outcrops form several continuous ridges and groups of small hills striking in a north-east to south-west direction. These consist mainly of schists dipping to 80° under a thick band of quartzite. The quartzites are reddish. The schists are reddish and greenish and traversed by quartz veins. These ancient crystalline and metamorphic rocks with gneisses and schists are generally covered by red sandy soils. Such soils though shallow on the hills are suitable for supporting forests of Dhok (Anogeissus pendula) and its associates.

The geology of the south-eastern region is mostly shared by these two rock systems. Their distribution is broadly determined by the two important faults which occur in this region about 12 miles apart and run in the north-east to south-west direction. The south-eastern fault known as "Great Boundary Fault of Rajputana" passes through Phalodi-Bodal-Lahpur-Chiroli and Quadeen. It is along this fault that Gwaliors are brought against Vindhyans. It also divides along its length and includes between its branches upper Vindhyan folded in anticline and brought against lower Vindhyan and Gwaliors. The north-western fault is not very clear owing to the large spreads of alluvium under which it lies but is a real fault and important in effects. At different places along its length the lower Vindhyans, Kaimur, lower Rewa and upper Bhanders are brought against Gwaliors. The lower Vindhyans are generally confined between these two faults. Apart from these faults of Bher, Dhundarmal Ka Darwaza, Anantpura and Berda are significant.

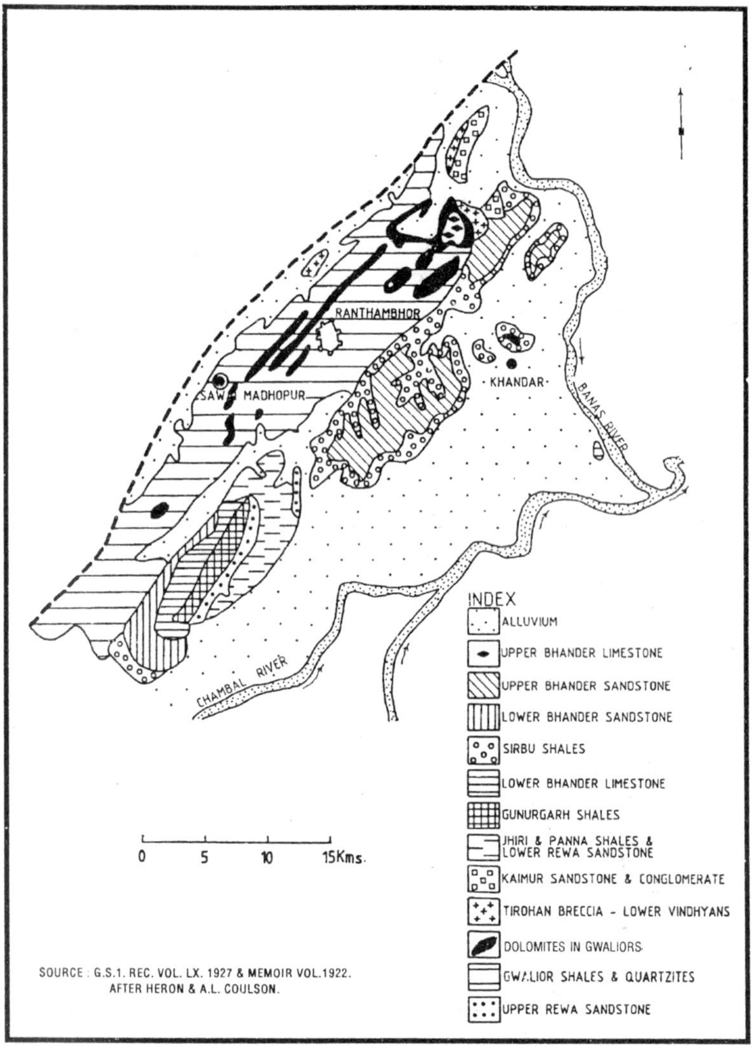

FIG. 2.7 : *Ranthambhore Sanctuary — Geology*

The Gwalior System

The rocks of this system are comparatively exposed over a large area north-west of the eastern fault and the Banas river. They are least exposed as syncline comprise chiefly of quartzite interbedded with subordinate shales and dolomites flows and sills. Quartzites are succeeded downwards by shales and slates and are pink or reddish and compact. Shales are dark red, purplish red or black or spotted in these colours. Dolomites are exposed in south-west of Anantpura, Kachida Valley near Berda, Lakarda areas of the Reserve and run along the bottom of deep narrow strike valleys. The soil on the rock of Gwaliors is finer clayey and shallow. The best soil resembling black cotton soil is found on the outcrops of dolomites supporting miscellaneous species specially Cheela, Raunj, Khair, Pipal etc.

Lower Vindhyans

They are exposed near Shyampura, Dhundarmal Ka Darwaza and Bher area of the Reserve containing quartzites and conglomerates.

Upper Vindhyans

They are exposed mainly between south-eastern fault and the Chambal river where upper Bhander sandstone with a horizontal strata forms large plateau lying over a mass of crumbling sirbushales; south-east of south-east fault and north-east of Dhundarmal-Ka-Darwaza where lower Rewa sandstone, Panna shales and Kaimur sandstone are exposed against lower Vindhyan breccia, sandstones or conglomerates.

The Vindhyan rocks here weather into fine sandy soils, aluminous, calcarious and ferrugeneous soils including deep dark loams and black soils which support miscellaneous species including Cheela, Tendu, Dhok etc.

Lineaments

Lineaments are defined as any linear features which are topological, structural and tectonic in origin. The lineament map (Fig. 2.13) prepared with the help of satellite data provides a distinct picture of lineaments distributions. The Aravalli ranges are affected by the criss-crossing linears whereas the Vindhyan ranges are totally free from them which clearly indicates the different ages in the evolution of the rocks and considerable time gap existed between the two different groups of rocks. Most of the lineaments end up only within Aravallis and do not continue into the adjacent plain on the west or horizontal Vindhyans on the east. Few linears seen or interpreted on the plain areas may be of different age and origin through which some present-day streams flow. So many vertical scarps and sudden elevations of Ranthambhore Fort from the Valley are due to the vertical movements along the lineaments.

The lineaments are in NW/SE, NE/SW and N/S direction. But the NW/SE lineaments are prominent and they are younger to NE/SW lineaments as they displace them. The fractured or jointed quartzites of Aravalli formation are displaced horizontally along the major lineaments which is running in NW/SE direction south of Kachida village (Fig. 2.13).

Drainage

The drainage of the Ranthambhore Sanctuary is flowing towards south and south-east. The streams and nallahs of the south-east drain into the Banas river and those of south into the Chambal river (Fig. 2.8). Most of the streams are controlled flowing along the weak lineament zone. The prominent being the Bagda Nalla, Padam and Rajbagh Talao.

Relief Features

The study of general relief and the total landscape is of paramount importance in an ecological study as they determine the distribution of plant communities and provide habitat to wild animals. A maze of low hills and parallel ranges form the general landscape of the area. These hills are of the Aravalli and Vindhyan mountain system which have entangled here. The general relief varies from 250 to over 450 metres altitude above mean sea level. The main cliff on which Ranthambhore fort is located ranges from 472 to 482 metres high above mean sea level. The ranges extend from south-west to north-east with highest culminations around 480 metres above mean sea level. A few isolated hills are on the north-east, on one such hill the famous Khandar fort exists. The steep faces of the hills and ranges are on the north-west which merge with the plains of Banas river. The northern slopes of ranges have steep escarpment which is a characteristic feature of the Vindhyan uplands. The south-eastern maze of hills with hogback crests and flattened tops have escarpments in the south-east direction but have narrow, deeply dissected valleys on the north-west slopes.

The entire hilly area presents a sensile topography which is highly

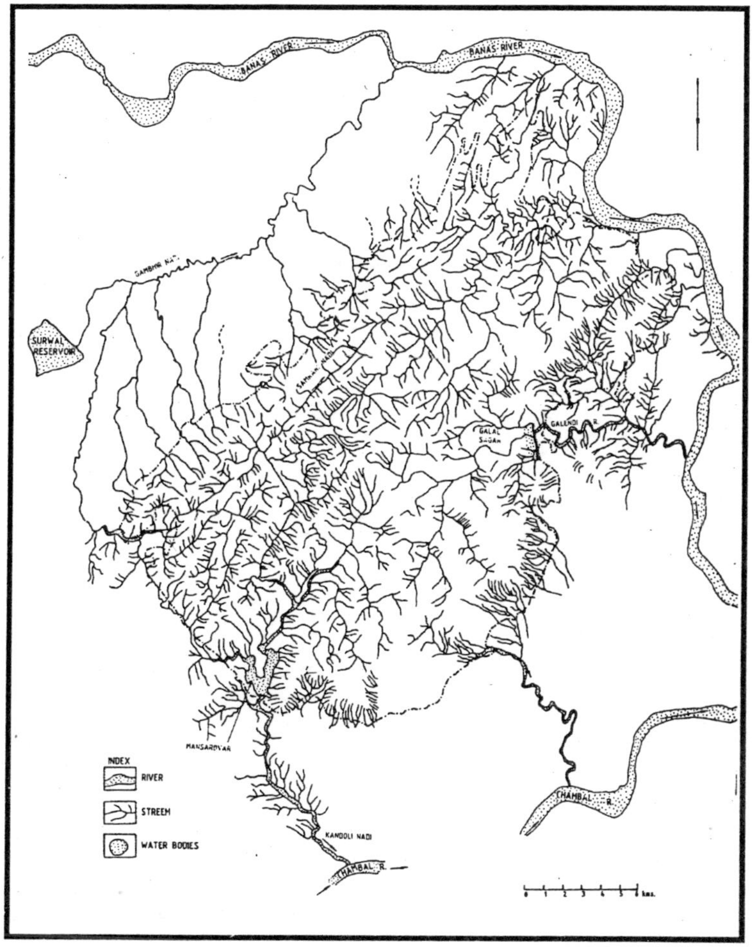

FIG. 2.8 : *Ranthambhore Sanctuary : Drainage*

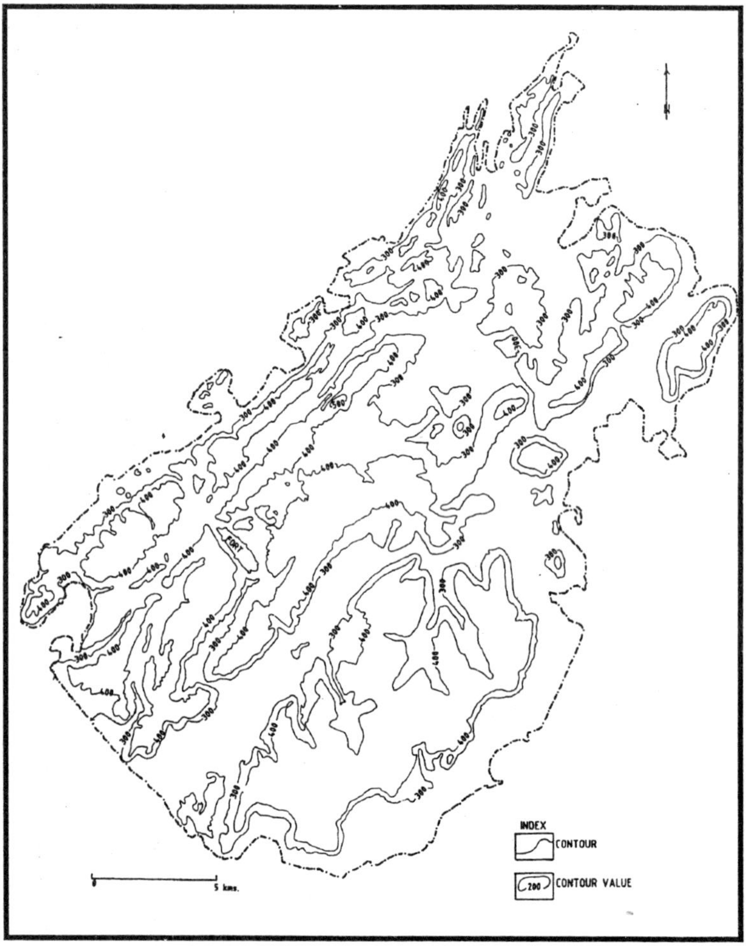

FIG. 2.9 : *Ranthambhore Sanctuary — Contour Map*

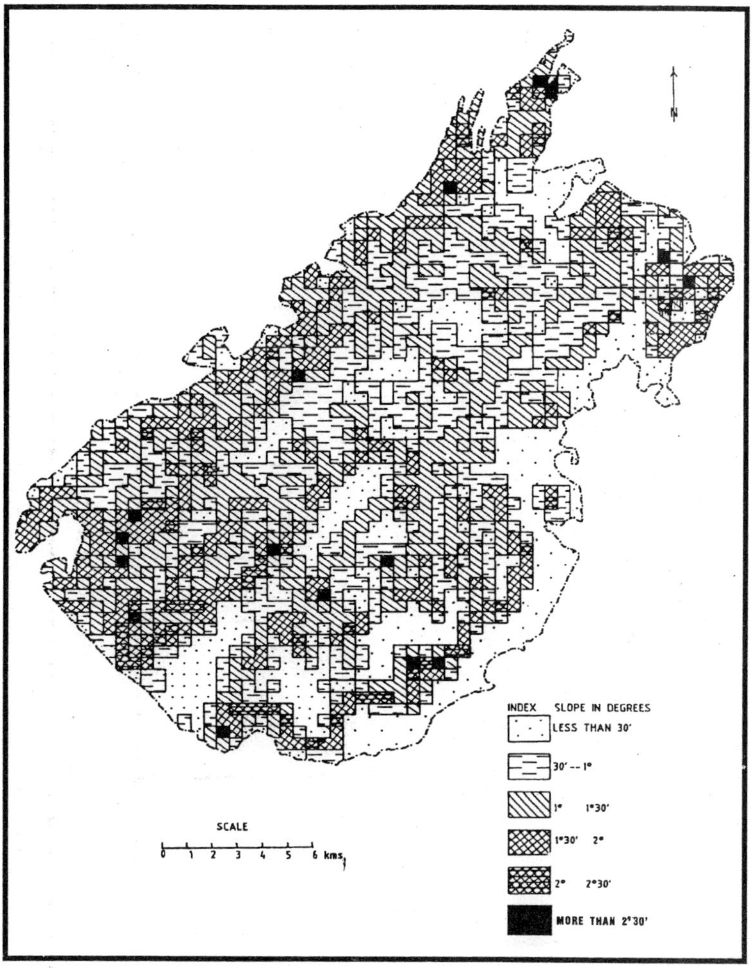

FIG. 2.10 : *Ranthambhore Sanctuary — Slopes*

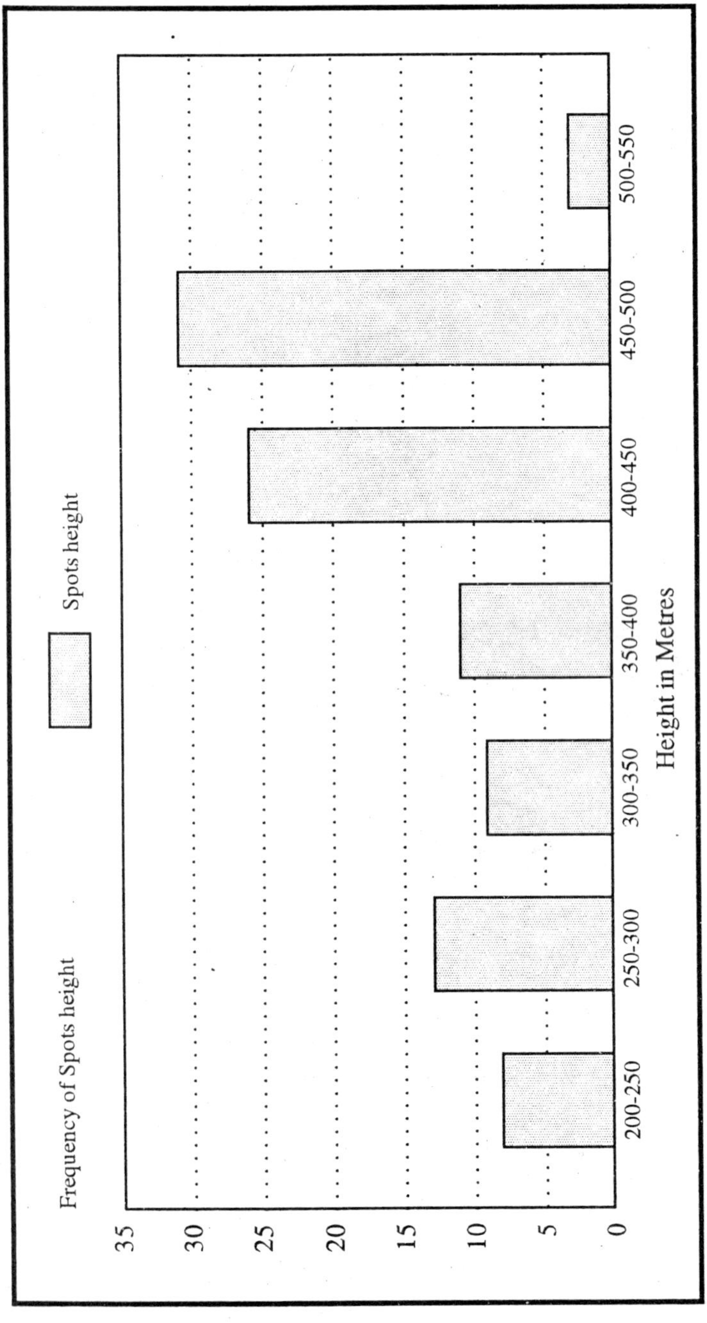

FIG. 2.11 : *Ranthambhore Sanctuary—Altimetric Frequency Histogram*

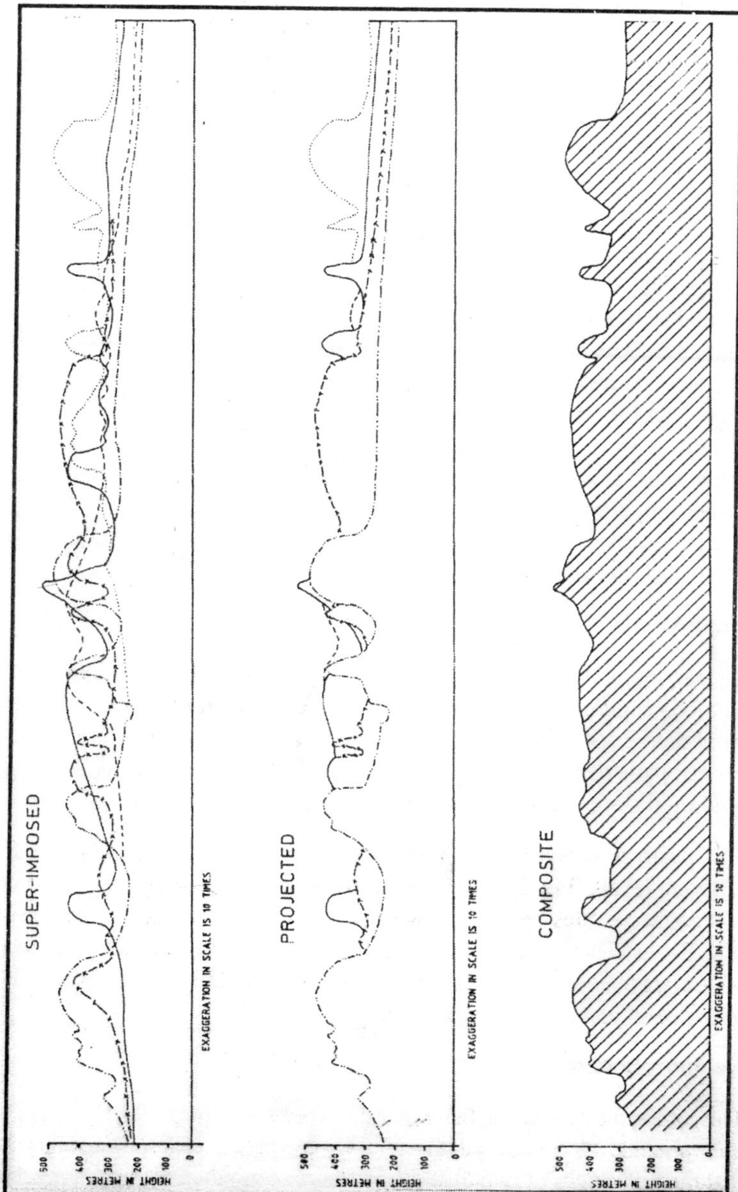

FIG. 2.12 : *Ranthambhore Sanctuary — Profiles*

dissected and rugged (Fig. 2.9). In the Sawai Madhopur block the Aravalli system prevails. Here, there are two main ranges called Salawai Dang and the Ranki Dang, both covered with fairly dense forest. To the south of these ranges are Rawanina Dungar hills separated from the main park by lowlands on which the villages of Sheopura, Halonda, Kheri, Bhagwanpura and Todra lie. As these villages lie in the buffer zone, they have been evacuated.

Most of the hills are of Vindhyan system, chiefly in the south-east, east and north. In the south, outside the limits of the park are a few villages like Chhan, Behraonda, Khurd, Phiriya and Mordoongri which abound with human activities. These villages are located in the flood zone of the Chambal river, in between the confluence zone of the Banas and the Chambal. In the east, lie the Mai Khurd, Mai Kalan, Talaora, Naipur and the historic Khandar village. The area in their immediate vicinity is an open scrubland with small patch of open mixed forest of Kikar trees. In the north, outside the precincts of the park are few villages such as Shyampura, Endwa, Basu, Doongri, Garhi and Sawata Kalan, all adjacent to the channel of the Banas and having small agricultural fields on the rich, fertile alluvial plains of this river.

Slope Analysis

An average slope map has also been prepared applying Wentworth's method. Figure 2.10 reveals that the average slope ranges from 0.30° to more than 2.30°. The high slopes are concentrated mainly in the north-eastern parts of the sanctuary. At the foot of plateau regions and valleys minimum slope below 1° is discernible.

Analysis of various profiles indicates presence of planer surface at different levels (Fig. 2.12). The results of superimposed, projected and composite profiles are also supported by altimetric frequency analysis (Fig. 2.11). The concentration of spot heights is at three levels i.e. above 400 metres, 300 metres to 450 metres and below 250 metres, indicating presence of planer surfaces at these heights.

Geomorphology

The topography varies from gentle slopes to steep vertical rocky escarpments. A unique geomorphological feature of this Sanctuary is that the hill ranges of two different systems of mountains meet along the Great Boundary Fault. The conical hillocks of the older Aravallis are juxtaposed with the flat topped hills of the Vindhyans. The altitudinal range is from 215 metres to 507 metres from mean sea level.

A detailed geomorphological map on 1:50,000 (Fig. 2.13) of the Ranthambhore Sanctuary and surrounding areas has been prepared on the basis of landsat data procured from the Institute of Sustainable Development Studies, Jaipur under IMSD project. Field checks have also been made in order to substantiate the information derived from landsat imageries.

The Ranthambhore Sanctuary is very unique geomorphologically and landforms have been developed by the combined action of endogenetic and exogenetic forces. The landforms of this region are fluvial, denudational and structural in origin. The detailed description of the various landforms is given below :

Fluvial Origin

The flood plains and valley fills are important forms of fluvial origin. Flood plains are confined along the Banas. Valley fills are few.

Denudational Origin

The geomorphic features of denudational origin are :

1. Denudational hills on quartzites.
2. Buried pediplains on slates/schists having less than 5 metres of mantle.
3. Buried pediplains on slates/schists having a regolith cover of 5 metres.
4. Denudation hills on brecciated rocks which are low to high hills of tirohan breccia.
5. Denudation hills on limestone.
6. Buried pediplains on shales — this is a flat to undulating plain carved out on the Vindhyan shales having shallow soil cover.
7. Buried pediplains on shales having flat to undulating plain carved out on the Vindhyan shales with 5 metres soil covers.
8. Pediplains on limestone.

Fig. 2.13 ***Contd.***

FIG. 2.13 : *Ranthambhore Sanctuary — Geomorphological Map*

Fig. 2.13 ***Contd. ...***

Map Symbol	Geomorphic Unit	Description	Ground water prospect
Fluvial Origin			
F P	Flood Plain	Recent deposits of gravel, sand, silt etc. along the margins of the streams unconsolidated quaternary sediment occupying narrow stretches within pediplain	Very good suitable for both shallow and deep aquifers
V F	Valley fill	Unconsolidated quaternary Sediment occupying narrow stretches within pediplain	Very good suitable for shallow aquifers
Denudational Origin			
DH - (Q)	Denudation hill on quartzite	Small, Isolated hills of quartzite	Poor
BPP - S(Sl)	Buried pediplain on slates/schists complex shallow	Nearly flat to undulating plain carved on slates/ schists association having less than 5 mts. of mantle	Moderate, suitable open dug-well
BPP - M(Sl.)	Buried pediplain on slate/schists complex-medium	Nearly flate to undulating plain on curved on slates/ schists association having a mantle of 5 mts. thickness	Good suitable for dug-wells and dug cum bore wells
DH - (Br)	Denudational hills on brecciated rocks	Low to medium high hills of Tirohan braccia with gentle slopes	Poor
DH - (LSt)	Denudational hill on limestone	Low lying hills of limestone along with the brecciated rocks	Poor
BPP - S(sh)	Buried pediplain on shales-shallow	Nearly flat to undulating plain carved on Vindhyan shales having shallow soil cover	Moderate suitable dug-wells
BPP - M(sh)	Buried pediplain on Vindhyan shales-Medium	Nearly flat to undulating plain carved on Vindhyan shales having 5-20 mts. soil cover	Good suitable for dug-wells and dug cum bore wells
PP - (Lst)	Pediplain on limestone	Flat bedded exposed limestone area gently slopping areas at the feet of plateaus and hills carved on different lithologies	Poor Generally poor except along Tirohan breccia which are suitable for deep aquifers

Structural Origins

SH - (Q)	Structural hills on quartzite	High rising strike ridges of quartzites with steep slopes	Poor
DPT - (Q)	Denudational Plateau on quartzite	A plateau-cuesta monoclinal hill complex on quartzite highly jointed and dissected	Poor Perched water possibilites in open joint valleys
SV - (sl)	Structural valley in slates	Narrow linear valleys in interbedded slates formed due to fractures/joints	Good Suitable. Only for shallow aquifers
SH - (Sst)	Structural hills on sandstones	Linear and arcuate ridges of low to medium dipping sandstone having flat tops	Poor
SH - (Lst)	Structural hills on limestone	Linear, narrow strike ridges of the Limestone interpredded with Sandstone and Shale scree	Poor
PT - (Sst)	Plateau on sandstone	High rising Sandstone capped plateau with interbedded Shale and Limestone	Poor
DPT - (Sst)	Dissected plateau on Sandstone	Highly dissected high rising plateau showing prominent gully erosion along joints and faults, bounded by escarpment and steep slopes	Moderate prospect along faults, fractures etc.
SV - (Sh)	Structural valley on Vindhyan shales	Valley formed over Vindhyan shales and interbedded Lime-stone as a result of faulting and folding	Good, suitable for shallow and deep aquifers

Map Symbols	Structural	Groundwater Origin	Topographic Symbol Prospects
Ravines (Ra)	Lineament	Excellent (E)	Geomorphic Contact
River bed with water	Fault	Very Good (VG)	
River bed without water	Fracture	Good (G)	
	Structural trends	Moderate (M)	
Scarps		Poor to nil (P)	

Structural Origin

1. Structural hills on quartzites having high rising strike ridges with steep slopes.

2. Denudational plateau on quartzites mainly confined to south-western part.

3. Structural valley in slates — a linear structural valley is discernible on map from Sawai Madhopur towards north-east of the Sanctuary.

4. Structural hills on sandstones — these are linear and arcuate ridges of low to medium dipping sandstones having flat tops. Such features are confined in the south-western part close to great boundary fault.

5. Structural hills on limestone.

6. Plateau on sandstone — the area south-east of great boundary fault is characterised by such a geomorphic feature. It is a prominent feature having steep scarps.

7. Dissected plateau on sandstone — it is a highly dissected high rising plateau showing prominent gully erosion along joints and faults bounded by escarpment and steep slopes.

8. Structural valley on Vindhyan shales — such valleys have been formed over the Vindhyan shales and interbedded limestones as a result of faulting and folding.

3

HUMAN AND ECONOMIC ENVIRONMENT

Population

THE population of the area around the sanctuary is par excellence agrarian, engaged in primary activities of agricultural production, animal husbandry and cottage industries which provide little scope to develop secondary and tertiary activities. Moreover, the only town of importance — Sawai Madhopur — is a sleepy town, which, except for its importance as a junction station for transhipment of goods and traffic to and from Jaipur city, is devoid of any major industry and business centre. (The only major industry was J.K. Cements, which has now closed down.) Under these conditions, primary occupations are to remain the mainstay of all the villages, with little opportunities for rural population to be engaged in non-agrarian sectors.

Population Density

Here, we take into account the density of the 54 villages, which exist on the fringe and peripheral areas of the sanctuary. The core area of the sanctuary is entirely free of any human population while the buffer zone villages portray a density of less than 100/km^2 which is normal, as these villagers have only substantive farming on the limited agricultural areas (Figs 3.1 & 3.2).

Population density ranging from 100 to 200/km^2 is however markedly noticeable in the north, both in Khandar and Sawai Madhopur blocks, mainly due to the hilly terrain, providing few means of land transport and an agricultural land which is under constant threat of inundation by the Banas river. But even this density is sufficiently higher as compared to Rajasthan average (129/km^2).

Density ranging from 200 to 300 is found in isolated pockets of north-east and also south of Sawai Madhopur town in some villages, which are agriculturally fertile and have better means of transport.

Higher densities ranging from 300 to 400 are found only in few villages, sporadically located apart as near Sawai Madhopur town and south of Man Sagar Lake. This may either be due to high fertility rates or on account of immigration from displaced persons of villages, formerly inside the sanctuary.

Densities of even high grade ranging from 400 to 500 and even more are found in Khandar village and to the north of Sawai Madhopur Town, which is due to immigration of population to these regional nodal centres in want of self-employment or to work in some trade.

Population Density Variations

Population is a dynamic factor and keeps changing, primarily due to the natural growth and further on account of the availability of basic land resources needed for sustenance and the provisions of infrastructure activities. The population variations of two decades viz. 1971-81 and 1981-91 computed for the villages of this region show glaring changes (Table 3.1).

Out of the 54 villages surveyed, two villages viz. Gopalpura and Kailashpuri are of recent origin and have been established in the present decade. These two villages constructed by the Forest Department house the displaced persons of the villages which were formerly inside the core area of the sanctuary. They are located at the south-west fringe of this sanctuary and their present population (1991 census) is 236 and 665 respectively.

The remaining 52 villages are of earlier origin and any deviations of population in them, besides the natural growth, are on account of voluntary emigration to the adjacent villages or due to the exodus of the people outside on account of scarcity of roads, food availability, lack of agricultural land and fear of the wild carnivores of the sanctuary which frequently roam about now.

In the decade 1971-81, 16 villages viz. Parli (49.61%), Badhlao

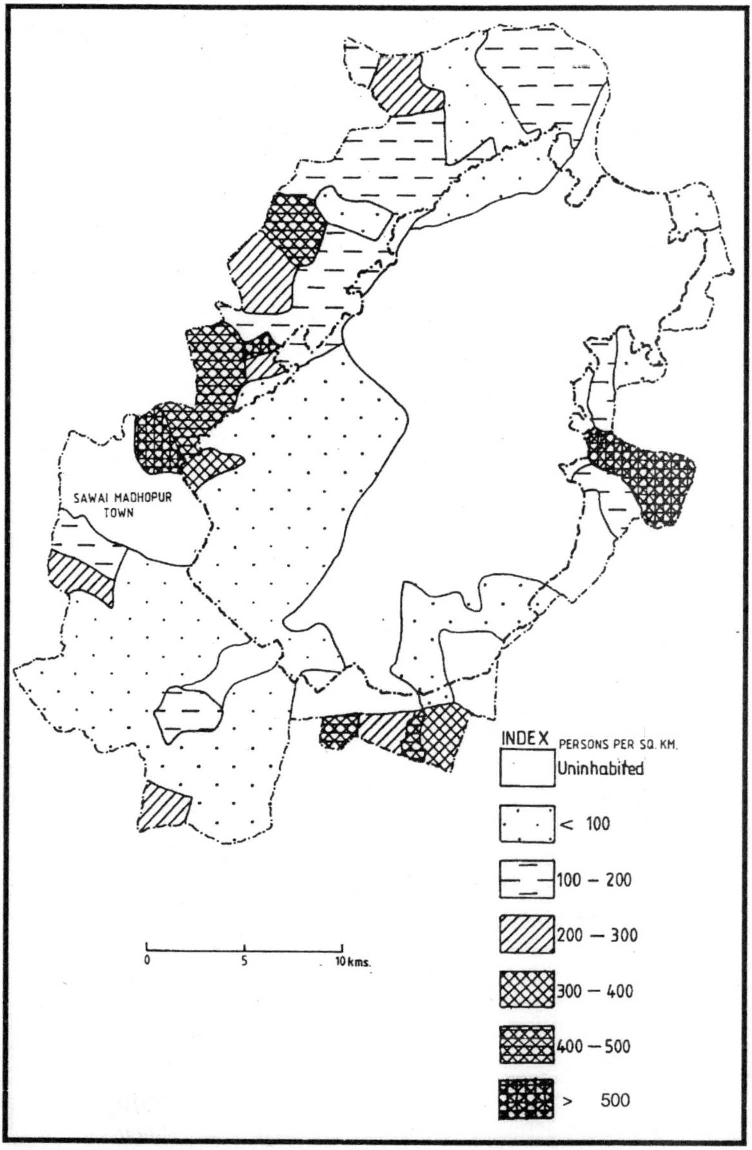

FIG 3.1 : *Ranthambhore Sanctuary — Population Density 1991*

TABLE 3.1 : **Density of Population**

S. No.	Village Name	1971			1981			1991			Variation 1981-91	Variations Density1981
		Area km^2	Population	Density	Area km^2	Population	Density	Area km^2	Population	Density		
1	2	3	4	5	6	7	8	9	10	11	12	13
1.	Olwara (Sąwai Madhopur)	15.20	953	62.70	15.17	1193	78.60	11.22	1207	107.60	290	36.90
2.	Endwa	69	1192	172.8	6.91	1406	203.5	6.98	17.28	247.6	44.1	21.7
3.	Shyampura	18.9	2541	134.4	19.15	3066	160.1	19.12	3810	199.2	39.1	24.4
4.	Parli	5.0	361	72.2	5.04	504	100.1	5.00	724	144.8	44.7	44.7
5.	Bhadlao	13.4	363	27.8	13.55	551	40.7	13.40	708	52.8	12.1	29.7
6.	Kundera	8.0	2495	311.9	8.08	3022	374.0	8.08	3544	437.5	63.5	17.0
7.	Rawal	8.6	1286	149.5	8.55	1596	186.7	8.65	2025	234.1	47.4	25.4
8.	Uliyana	9.5	638	67.2	9.42	803	85.2	9.58	1070	111.6	26.4	31.0
9.	Khandoj	0.8	43	53.8	1.04	115	110.6	1.03	133	129.1	18.5	16.7
10.	Sherpur	3.3	364	110.3	3.36	548	163.1	3.36	921	274.1	111.0	68.1
11.	Kutalpura Maliyan	1.0	297	297.0	1.02	393	385.3	1.01	524	513.7	128.4	33.3
12.	Khawa	3.6	249	69.2	3.64	291	79.9	3.60	429	119.2	39.3	49.2
13.	Jamool Khera	5.3	508	95.8	10.12	626	61.9	5.30	822	154.8	929.0	150.0
14.	Khilchipur	11.8	3176	269.2	11.95	4123	345.0	11.95	5019	420.0	75.0	21.7
15.	Ramsinghpura	3.2	865	270.3	3.28	1227	374.1	3.27	1555	475.5	101.4	27.1

(Contd.)

TABLE 3.1 : *Contd.*

1	2	3	4	5	6	7	8	9	10	11	12	13
16.	Khatipura	2.2	835	379.5	2.19	1103	503.7	2.19	1516	719.6	215.9	42.9
17.	Borif	3.9	569	145.4	3.96	698	176.3	3.97	1014	255.4	791.0	449.0
18.	Madhosinghpura	1.1	239	217.3	1.15	342	297.4	1.15	445	386.9	89.5	30.1
19.	Nimli Khurd	15.8	411	26.0	15.99	575	36.0	15.99	762	47.7	11.7	32.5
20.	Nimli Kalan	16.4	419	25.5	16.54	653	39.5	16.64	903	54.3	14.8	37.5
21.	Kalibhat	5.0	232	46.4	5.00	95	19.0	5.03	83	15.7	-3.3	-17.4
22.	Kushalipura	6.7	116	17.3	7.05	204	28.9	6.80	238	35.0	6.1	21.1
23.	Hajjam Kheri	4.8	67	14.0	5.04	246	48.8	4.87	226	46.4	-2.4	-4.9
24.	Halonda	8.3	459	55.3	8.86	623	70.3	8.39	879	104.8	34.5	49.1
25.	Mordoongri	5.5	151	27.5	5.73	232	40.5	5.49	298	54.3	13.8	34.1
26.	Bodal	17.0	236	13.9	17.84	322	18.0	17.09	443	25.9	79.0	43.9
27.	Bheronpura	23.6	205	8.7	23.94	138	5.8	23.93	159	6.6	0.8	13.8
28.	Dumoda	16.1	634	39.4	16.91	1220	72.1	3.89	1002	257.6	185.5	25.73
29.	Baso Kalan (Khandar)	5.8	341	58.8	5.77	422	73.1	5.84	548	93.8	20.7	28.3
30.	Baso Khurd	7.5	418	55.7	7.55	559	74.0	7.54	596	79.0	5.0	6.8
31.	Bhoori Pahari	22.2	1331	60.0	22.50	1815	80.7	22.49	2304	102.4	21.7	26.9
32.	Dungri	11.0	1339	121.7	11.07	1190	107.5	11.07	1515	136.9	29.4	27.3
33.	Sawtan	12.5	868	69.4	12.60	734	58.0	12.60	94	74.6	1.6	28.6
34.	Kathooli Dhanapala	8.2	174	21.2	8.45	200	23.7	8.28	259	31.3	7.6	32.1

(*Contd.*)

TABLE 3.1 : *Contd.*

1	2	3	4	5	6	7	8	9	10	11	12	13
35.	Padra Nizamat	11.4	161	14.1	11.52	204	17.7	11.55	229	19.8	2.1	11.9
36.	Goth Bihari	7.0	467	66.7	7.06	635	89.9	7.06	872	123.5	33.6	37.4
37.	Sanbar	3.8	73	19.2	5.75	253	44.0	5.75	314	54.6	10.6	24.1
38.	Khandar	16.9	4324	255.9	14.10	5856	415.3	14.10	7361	522.1	106.8	25.7
39.	Rawrah	5.0	314	62.8	5.05	425	84.6	5.05	578	114.5	29.9	35.3
40.	Jaisinghpura	5.1	43.4	85.0	5.22	537	102.9	5.22	687	131.6	28.7	27.9
41.	Mei Kalan	20.5	958	46.7	18.39	1086	59.1	18.61	1365	73.3	14.2	24.4
42.	Indala	10.0	32	3.2	10.15	46	0.04	10.15	424	41.8	41.8	1045.0
43.	Jetpur	2.5	4.7	162.8	2.51	700	279.5	2.56	1049	409.8	130.3	46.6
44.	Chhan	19.6	2098	107.0	19.84	3596	130.8	19.85	4502	226.8	96.0	73.4
45.	Allapur	2.6	819	315.0	2.58	948	367.4	2.60	120.2	462.3	94.9	25.8
46.	Bahrawada Khurd	10.1	2246	222.4	10.23	3025	295.7	10.23	3929	384.1	88.4	29.9
47.	Niwandi (S.M.)	8.0	308	38.5	7.7	510	66.2	1.8	658	84.3	18.1	27.3
48.	Bhundeda	1.6	491	306.9	1.6	621	388.1	1.6	843	526.9	138.8	35.8
49.	Jeenapur	4.4	397	90.2	44	497	112.9	4.5	681	151.7	38.8	34.4
50.	Himmatpur	0.3	109	363.0	0.3	166	553.3	0.3	197	579.4	26.1	4.7
51.	Qila Ranthambhore	104.3	4.3	0.41	105.5	668	0.64	1054	75	0.71	.07	10.93
52.	Sakhwas	1.3	249	191.5	1.3	276	212.3	1.3	494	389.0	176.7	83.2
53.	Gopalpura (Khandar)	—	—	—	—	—	—	9.5	236	24.8	24.8	NV
54.	Kailashpuri (S.M.)	—	—	—	—	—	—	12.3	665	54.1	54.1	NV

(51.79%), Khandoj (167.44%), Sherpur (49.72%), Ramsinghpura (41.84%), Madhosinghpura (43.09%), Nimli Khurd (39.90%), Nimli Kalan (55.64%), Hajjam Kheri (267.16%), Mordoongri (53.64%), Dumoda (92.42%), Sanbar (246.57%), Indala (43.75%), Chhan (71.40%), Niwandi (65.6%) and Qila Ranthambhore (52.3%) showed an extraordinary growth of population beyond the natural increase. It was due to the immigration of those people who were formerly living inside the sanctuary and had been asked to vacate their villages so as to grant this sanctuary the status of a national park, where all human activities are totally banned and the area should be without any human habitation. So we find that in this decade, the population of a few villages like Kalibhat (–59.05%), Bheronpura (–32.68%), Doongri (–11.12%), Sawtan (–14.45%) has substantially decreased.

In the decade 1981-91, again there has been an abnormal rise of population in some villages and a sharp fall in others. The abnormal rise is noticeable in Parli (43.65%), Sherpur (68.06%), Khawa (47.42%), Khatipura (42.88%), Borif (45.27%), Halonda (41.09%) and Indala (821.73%), the last one showing the highest increase. A field visit in this regard shows that today in 1998, this village has hardly a hundred people but in 1991 at the time of census, Indala had 424 persons. It is likely, therefore, that initially at the beginning of the present decade, Indala must have had those persons who were evacuated from the Park precincts and later on shifted to the new villages of Kailashpuri and Gopalpura.

During the last decade villages again showed a negative growth. These are Kalibhat (–12.63%), Hajjam Kheri (–8.13%) and Dumoda (-17.6%). Dumoda is one of the largest villages with a population of 1002 (1991 census) and slow emigration indicates that due to squeezing civic facilities in this area, the villagers are staggeringly moving out.

Variations in populations of the last decade exhibit drastic changes of over 100% only in 4 villages, sporadically located — Indala and Chhan in the south showing the highest variation, Sherpur in the north-west and Kailashpuri — the new village — in the south-west (Fig. 3.2).

Scheduled Caste and Scheduled Tribe Population

The population of these two groups is astoundingly large in this area (Table 3.2). Out of the 54 villages surveyed, 21 villages have a large scheduled caste population which is beyond the Rajasthan average (Table 3.2). Out of these 21 villages, a few have much higher percentage like

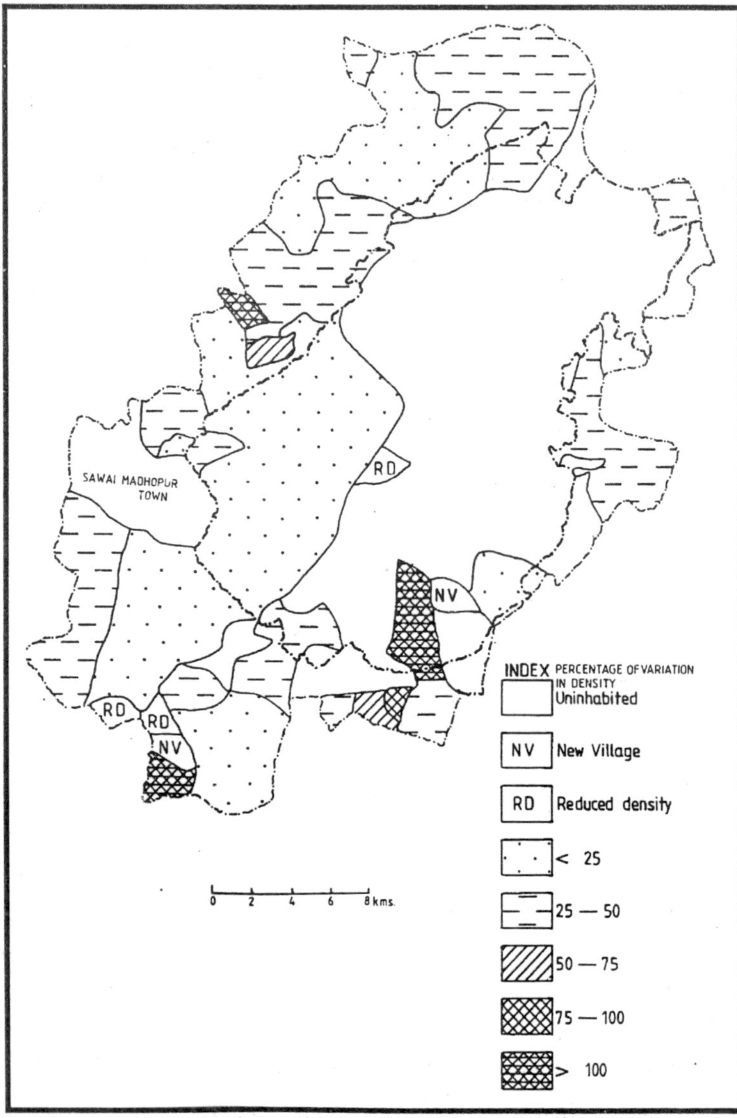

FIG. 3.2 : *Ranthambhore Sanctuary — Variation of Population Density 1981-91*

TABLE 3.2 : **Percentage of SC and ST Population**

S. No.	Villages	ST	SC	S. No.	Villages	ST	SC
1.	Olwara (S. M.)	55.34	11.34	28.	Dumoda	—	10.07
2.	Endwa	63.42	8.50	29.	Baso Kalan (Khandar)	69.34	—
3.	Shyampura	65.27	9.29	30.	Baso Khurd	78.18	1.34
4.	Parli	6.62	27.34	31.	Bhoori Pahari	39.01	23.61
5.	Bhadlao	24.29	8.89	32.	Dungri	53.33	17.75
6.	Kundera	0.08	19.89	33.	Sawtan	—	27.69
7.	Rawal	69.72	4.49	34.	Kathooli Dhanapala	—	—
8.	Uliyana	79.25	1.68	35.	Padra Nizamat	—	23.14
9.	Khandoj	53.38	—	36.	Goth Bihari	—	—
10.	Sherpur	1.95	3.36	37.	Sanbar	0.63	12.42
11.	Kutalpura Maliyan	—	15.26	38.	Khandar	4.42	21.36
12.	Khawa	47.48	—	39.	Rawrah	—	95.32
13.	Jamool Khera	50.72	—	40.	Jaisinghpura	—	59.24
14.	Khilchipur	21.91	23.45	41.	Mei Kalan	—	19.78
15.	Ramsinghpura	0.12	46.49	42.	Indala	—	100.00
16.	Khatipura	3.93	43.84	43.	Jetpur	—	3.52
17.	Borif	48.32	36.39	44.	Chhan	0.26	23.58
18.	Madhosinghpura	0.22	52.80	45.	Allapur	0.83	17.63
19.	Nimli Khurd	61.54	3.01	46.	Bahrawada Khurd	0.25	35.42
20.	Nimli Kalan	6.97	30.18	47.	Niwandi (S.M.)	28.7	10.0
21.	Kalibhat	54.21	—	48.	Bhundeda	—	24.9
22.	Kushalipura	—	0.42	49.	Jeenapur	86.5	1.02
23.	Hajjam Kheri	73.89	—	50.	Himmatpur	—	100.00
24.	Halonda	0.11	79.63	51.	Qila Ranthambhore	5.3	17.3
25.	Mordoongri	—	30.87	52.	Sakhwas	—	8.5
26.	Bodal	9.70	9.02	53.	Gopalpura	- 0%	- 0%
27.	Bheronpura	—	—	54.	Kailashpuri (S.M.)	- 0	0

Indala (100.0%), Rawrah (95.32%), Halonda (79.63%), Madhosinghpura (52.80%), and Ramsinghpura (56.49%).

Concentration of scheduled caste population is noticeable along the fringe of the buffer zone of this sanctuary. Poor as they are, their sole occupations are firewood-selling and livestock-rearing, both of which can be easily pursued in the fringe zone both by legal and illegal methods (Fig. 3.3). Elsewhere the SC population is sporadic, below 20% or even below 10% which is normal.

The scheduled tribe population, mostly of the 'Meena' tribe, is glaringly larger in almost every village, showing thereby that this whole area has a predominance of this tribe. Meenas are mostly agriculturists but are also engaged in other activities like trade, transport, cottage and small-scale industries and official jobs. Fifty years of independence has given them ample time to seek official jobs and even higher administrative posts in the state and centre and so today, as a clan they are not oppressed to live below poverty line; rather they are the creamy layer of the rural society.

The highest percentages of scheduled tribes are found in Jeenapur (85.6%), Baso Khurd (78.18%), Uliyana (79.25%), Rawal (69.72%), Baso Kalan (69.34%), Shyampura (65.27%), Endwa (63.42%), Khandoj (53.38%), Nimli Khurd (61.54%), Olwara (55.34%) and Kalibhat (54.21%), Elsewhere, their population is less than 50% (Fig. 3.4).

It is noticeable that nearly 40% of the villages have less than 20% of the scheduled caste and tribal population, a little more than 12% villages have 20% to 40% scheduled caste population, about 5% have 40% to 60% of ST and SC population, about 9% have 60% to 80% ST/SC population and only 5% or less have more than 80% ST/SC population (Fig. 3.5). The preponderance of these people in these areas is coaxially related to the past history, when 'Meenas' dominated the 'Dang' area of erstwhile Jaipur state. The hilly, ravinous and eroded lands of these mountainous areas needed ample labour to till the soil and so these low caste people inhabited the areas where production was little. But today with the development of agriculture, provision of electricity on the wells and tubewells, availability of hybrid seeds and fertilizers, these areas have rich harvest of wheat, mustard and groundnuts, which fetch cash money to the peasants.

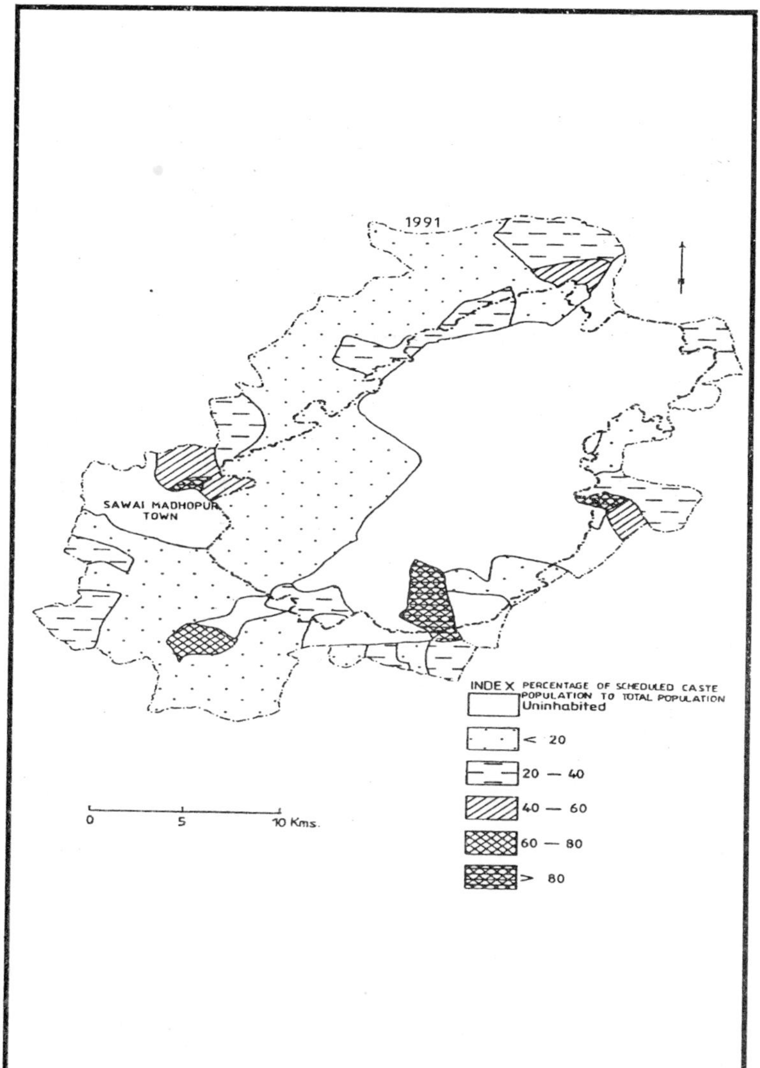

FIG. 3.3 : *Ranthambhore Sanctuary — Distribution of S.C. Population 1991*

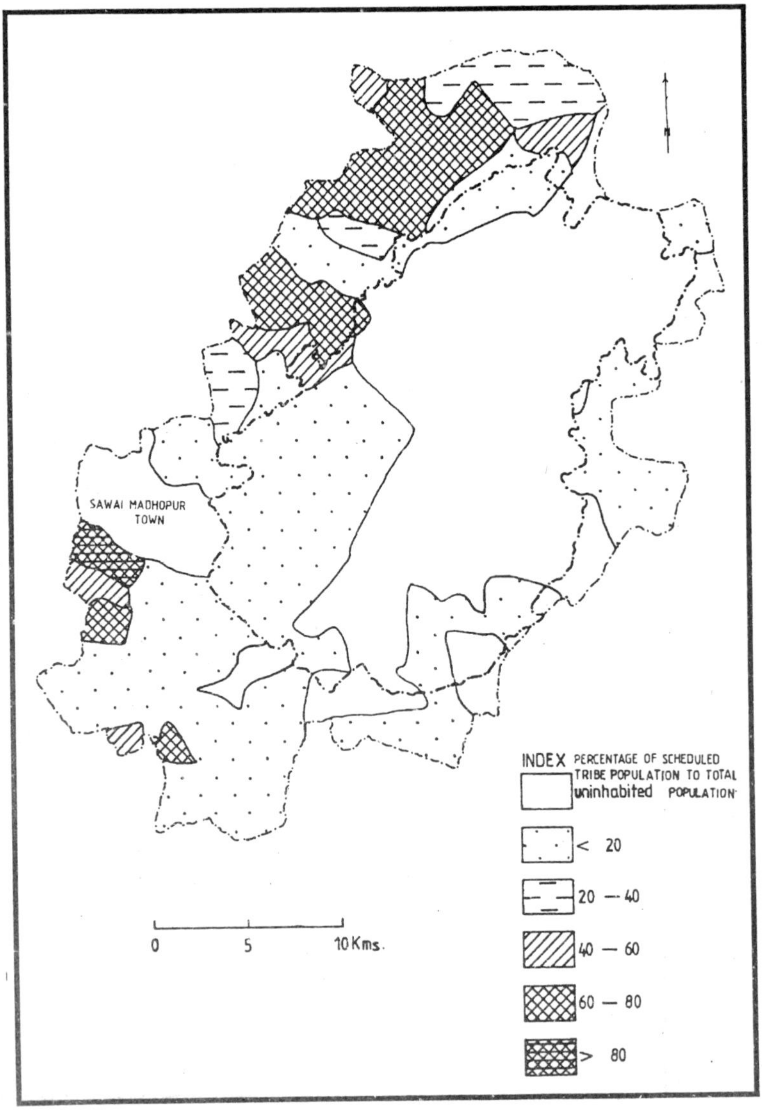

FIG. 3.4 : *Ranthambhore Sanctuary — Distribution of S.T. Population 1991*

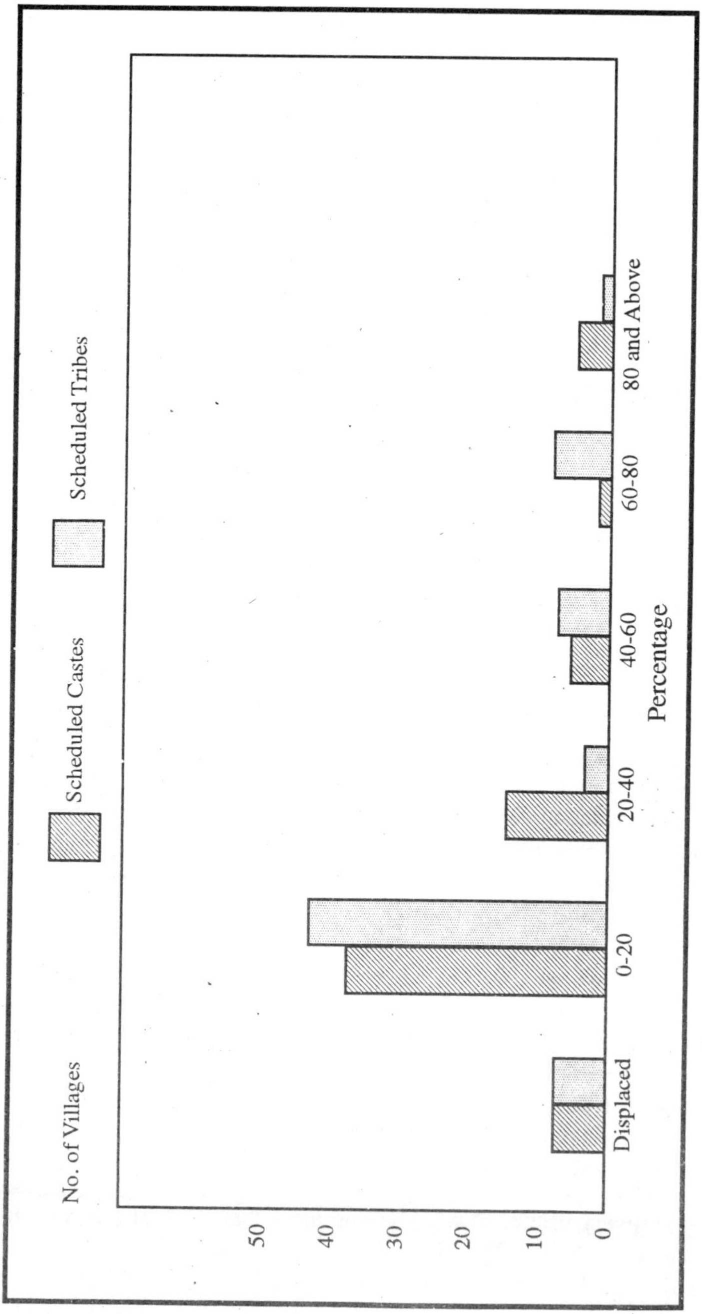

FIG. 3.5 : *Ranthambhore Sanctuary & Environs — SC & ST Population*

Population Engaged in Primary Occupations

The occupational data of 1991 census show that the occupational status of the people of these 54 surveyed villages is highly varied. Those engaged in primary occupations vary from 2.1% (Shyampura) to 53.3% (Baso Kalan). Availability of fertile land, proximity to surface and sub-surface water resources for irrigation and soil-moisture, plain vegetative lands for animal grazing and long distance from towns and market centres encourage agricultural pursuits.

Thus we find that 19 villages viz. Olwara (33.1%), Parli (45.9%), Bhadlao (32.9%), Viyana (42.1%), Handoj (47.4%), Khawa (46.6%), Borif (38.7%), Nimli Khurd (42.8%), Baso Kalan (53.3%), Bhoora Pahari (35.7%), Sawtan (49.7%), Kathooli Dhanapala (39.8%), Padra Nizamat (37.1%), Goth Bihari (49.0%), Indala (48.1%), Jetpur (30.7%), Chhan (32.7%), Allapur (49.7%) and Kailashpuri (32.3%) have more than 30% of population engaged in agricultural primary occupations and are primarily agrarian. Incidentally, these villages have the highest populations also, substantiating the view that larger population concentrations in Rajasthan, as also in India, are found in agrarian societies. The soil of this region, largely composed of recent and older alluvium of the Banas river, is productive and is therefore supportive to agrarian pursuits (Table 3.3).

Contrarily, quite a few villages have very low percentage of population engaged in farming. Thus Shyampura (21.1%), Sherpur (11.9%), Khatipura (5.5%), Bodal (8.8%), Khandar (11.8%), Bhundeda (10.7%), Jeenapur (12.0%) and Himmatpur (9.6%) have very low percentages of population in primary activities. The people of these villages, as the survey shows, are largely engaged in trade, cottage industries of 'Khas' and wooden toys, equipments, transport and administrative activities of the main town Sawai Madhopur. All the above cited villages have close proximity to metalled roads, village roads, railway stations, tourist resorts etc. Tourism, as we know, today is one of the major industries of Ranthambhore sanctuary and so quite a large number of people are engaged in plying of jeeps, maintenance of autovehicles, tourist hotels and resorts, selling merchandise and other gift items and other ancillary industries.

Primary occupations occupying more than 30% of the population are concentrated in the north, where the river Banas meanders to take a south and south-easterly direction, as rich alluvial soil suitable for

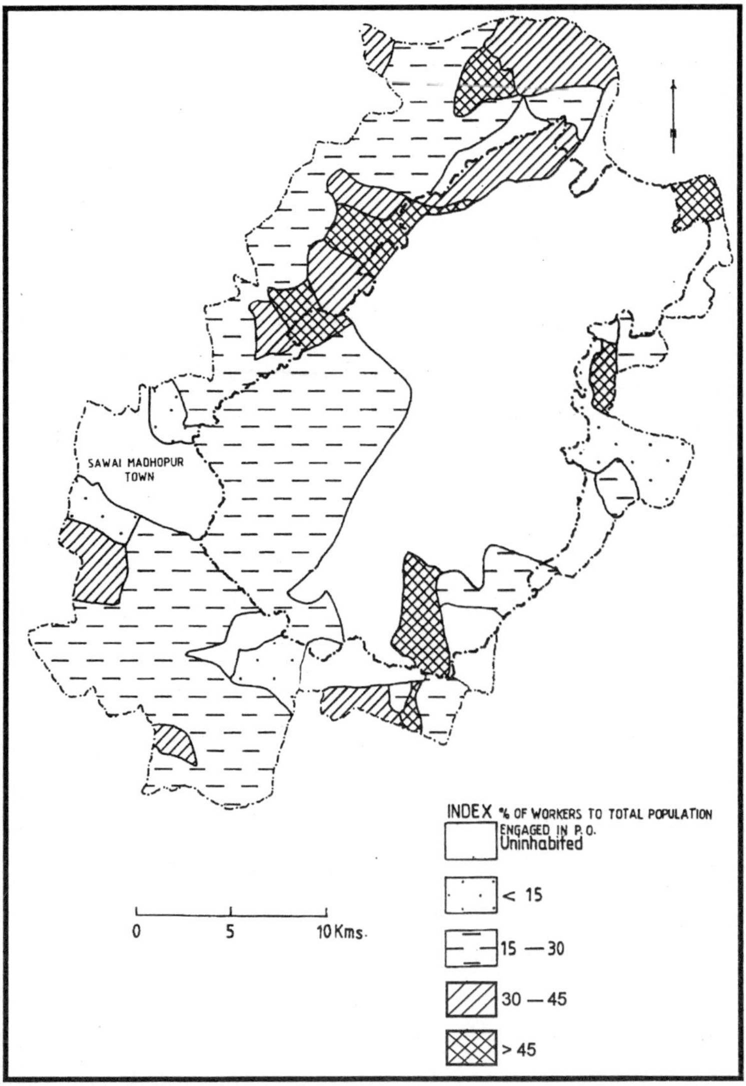

FIG. 3.6 : *Ranthambhore Sanctuary — Population Engaged in Primary Occupation 1991*

TABLE 3.3 : **Population Engaged in Primary Occupation (1991)**

S. No.	Village Name	Persons engaged in primary occu-pations	Total Popu-lation 1991	%age	S. No.	Village Name	Persons engaged in primary occu-pations	Total Popu-lation 1991	%age	S. No.	Village Name	Persons engaged in primary occu-pations	Total Popu-lation 1991	%age
1.	Olwara (S.M.)	399	1207	33.1	21.	Kalibhat	21	83	25.3	39.	Rawrah	84	578	14.5
2.	Endwa	364	1728	21.1	22.	Kushalipura	47	238	19.7	40.	Jaisinghpura	153	687	22.3
3.	Shyampura	842	3810	2.1	23.	Hajjam Kheri	67	226	29.6	41.	Mei Kalan	341	1365	25.0
4.	Parli	332	724	45.9	24.	Halonda	196	879	22.3	42.	Indala	204	424	48.1
5.	Bhadlao	233	708	32.9	25.	Mordoongri	83	298	27.8	43.	Jetpur	322	1049	30.7
6.	Kundera	698	3544	19.7	26.	Bodal	39	443	8.8	44.	Chhan	1463	4502	32.5
7.	Rawal	567	2025	28.0	27.	Bheronpura	47	159	29.6	45.	Allapur	597	1202	49.7
8.	Uliyana	451	1070	42.1	28.	Dumoda	267	1002	26.6	46.	Bahrawada	912	3929	23.2
9.	Handoj	63	133	47.4	29.	Baso Kalan	292	548	53.3		Khurd			
10.	Sherpur	110	921	11.9		(Khandar)				47.	Niwandi	134	658	20.4
11.	Kutalpur Maliyan	78	524	14.9	30.	Baso Khurd	165	596	27.7		(S.M.)			
12.	Khawa	200	429	46.6	31.	Bhoori Pahari	822	2304	35.7	48.	Bhundeda	90	843	10.7
13.	Jamool Khera	158	822	19.2	32.	Dungri	385	1515	25.4	49.	Jeenapur	82	681	12.0
14.	Khilchipur	799	5019	15.9	33.	Sawtan	467	940	49.7	50.	Himmatpur	19	197	9.6
15.	Ramsinghpura	274	1555	17.6	34.	Kathooli	103	259	39.8	51.	Qila	12	75	16.0
16.	Khatipura	87	576	5.5		Dhanapala					Ranthambhore			
17.	Borif	392	1014	38.7	35.	Padra Nizamat	85	229	37.1	52.	Sakhwas	127	494	25.7
18.	Madhosinghpura	83	445	18.7	36.	Goth Bihari	427	872	49.0	53.	Gopalpura	66	236	28.0
19.	Nimli Khurd	326	762	42.8	37.	Sanbar	77	314	22.9	54.	Kailashpuri	215	665	32.3
20.	Nimli Kalan	218	903	24.1	38.	Khandar	867	7361	11.8		(S.M.)			

cultivation is available in these narrow plains. On the slopes of the northern and north-western hills, rich pasture lands exist which are suitable for livestock-rearing, one of the most favoured agrarian occupations of this area (Fig. 3.6). Elsewhere primary activities are confined to the isolated villages near Khandar in the east, Inḍala in the south, Kushtala in the west and smaller areas in the south near Man Sagar.

An analysis of the occupational structure of all the 54 villages shows that the main workers are 32% and the marginal workers are very few (7%) showing that the dependency rate of population is very high. About 60% of the population is of non-working (Fig. 3.7). This indicates that the number of children and older people who cannot work for earning livelihood is very high, bringing in turn the poor standards of living. Most of this 'Dang' area is therefore infested with poverty and illiteracy, a factor which was latent so far. It has been only in the past two years that the state government has planned to develop the 'Dangs' both in the agrarian and educational sectors but the efforts have been half-hearted due to lack of funds which have to be pumped in to restore the economy.

Among the working population cultivators are predominant (67%) followed by agricultural labourers (11%) showing an ultra-dependency upon the soil and water resources (Fig. 3.8). Landless labourers are in large number which is a stigma to the region as these people either work on low wages and that too for a part of the year when sowing and harvesting seasons begin or they migrate to towns and cities to become permanent migrants and commuters, only to add to the population of urban centres which are already swelling and under constant stress.

Population engaged solely in livestock-rearing (7%), trade and commerce (4%), building construction works (3%) and other services (8%) forms a trifle minority and demands extension of job opportunities in the secondary and tertiary sector.

Decadal Growth of Population

The decadal growth of population plotted on scatter diagram for two decades 1971-81 and 1981-91 shows the trends of growth of the last two decades for the 54 villages taken in survey. Marked changes are discernible in the two graphs which foretell the wavering patterns of growth in individual villages, the reasons for which may be varied (Figs. 3.9 & 3.10).

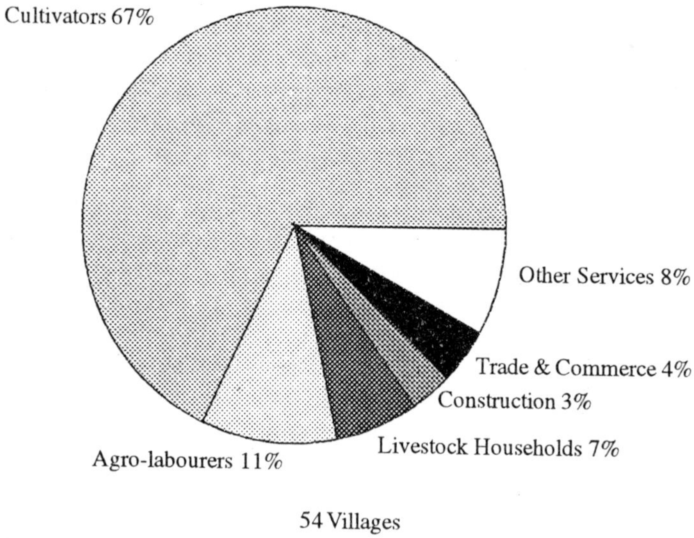

FIG. 3.7 : *Ranthambhore Sanctuary & Environs — Occupational Structure*

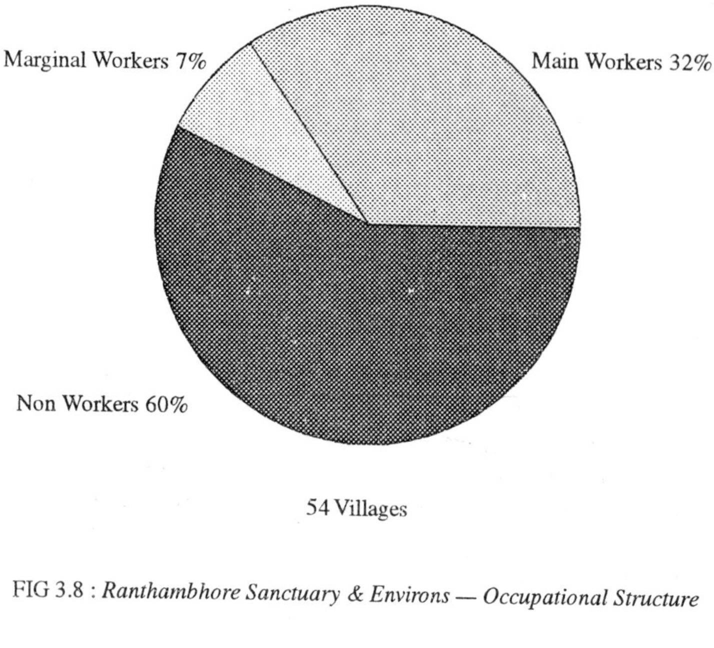

FIG 3.8 : *Ranthambhore Sanctuary & Environs — Occupational Structure*

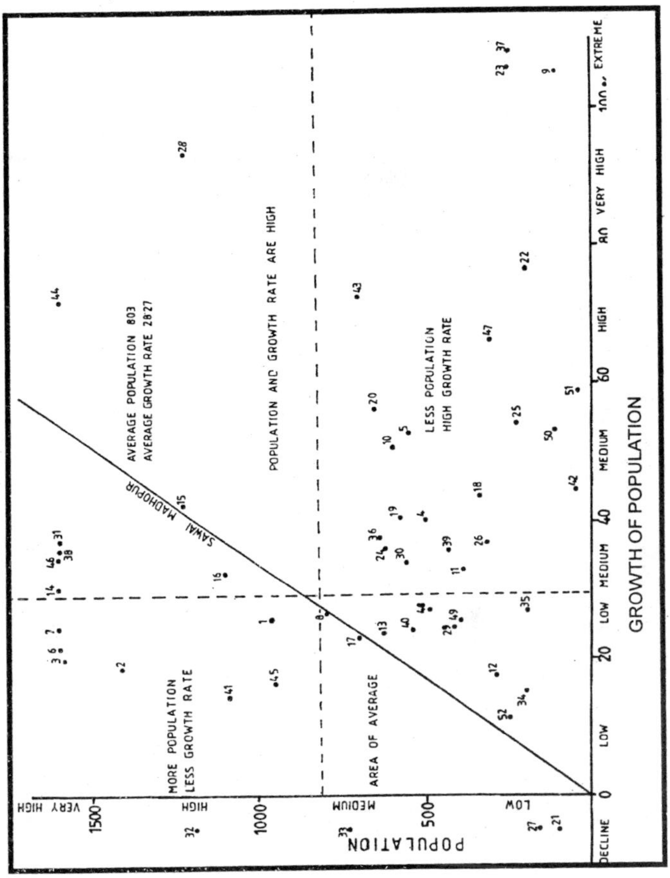

FIG. 3.9 : *Ranthambhore Sanctuary : Decadal Growth of Population 1971-81*

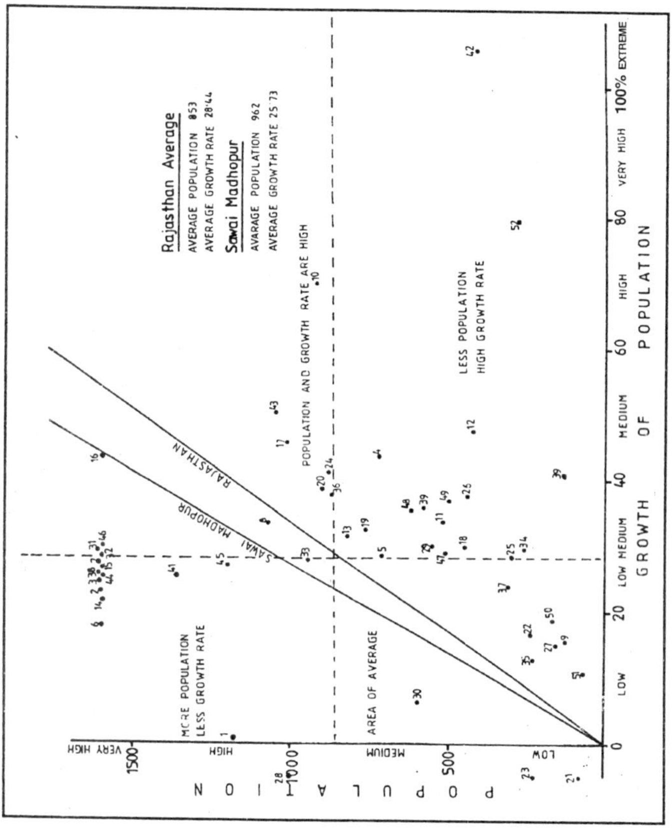

FIG. 3.10 : *Ranthambhore Sanctuary — Decadal Growth of Population 1981-91*

The analytical study of Fig. 3.9 drawn for 1971-81 decade to derive interrelation with the absolute population of the village and its growth rate in population shows 4 distinct sectors when the growth rate of Sawai Madhopur district is taken as a median line. These 4 sectors are:

(i) *Average Villages* — They have low to medium population (up to 800 persons) and low-medium population growth rate (30%) which was normal to Sawai Madhopur district. In this group lie 10 villages viz. Viyana (8), Khawa (12), Jamool Khera (13), Borif (17), Kathooli Dhanapala (34), Padra Nizamat (35), Jaisinghpura (40), Bhundeda (48), Jeenapur (49) and Sakhwas (52). (Note: Figures in parentheses show the village code of Table 3.4)

(ii) *High Population and Less Growth Rate Villages* — These villages have high population but recorded lower population growth rate (less than 30%). There are 7 villages in this category viz. Olwara S.M. (1), Endwa (2), Shyampura (3), Kundera (6), Rawal (7), Mei Kalan (41) and Allapur (45).

(iii) *Less Population and High Growth Rate Villages* — These villages have less population than the district average (803) but a higher growth rate than the district average (28.77%). Quite a large cluster of villages fall in this group but most notable among them are Khandoj (167.44%), Hajjam Kheri (267.16%) and Sanbar (246.57%). Truly, this is not a natural growth of population and speaks of the immigration of migrants who have been displaced from the villages formerly located inside the core area of the sanctuary.

(iv) *High Population and High Growth Rate Villages* — Only 2 villages fall in this category. These are Dumoda (28) which had a population of 1220 in 1981 and recorded a growth rate of 92.42% and Chhan (44) which had a population of 3596 in 1981 and a growth rate of 71.40%.

Moving ahead to the decade 1981-91, noticeable changes are discernible in the positions of these villages (Fig. 3.10). For example the villages with high population and high growth rate in this decade are — Sherpur (10), Borif (17), Nimli Kalan (20), Halonda (24) and Jetpur (43). The former villages of Dumoda (28) and Chhan (44) have now shifted to the left in the less growth zone.

Similarly, there has been an exorbitant increase in the number of

TABLE 3.4 : **Decadal Growth of Population**

(In Percentage)

S. No.	Village Name	1971-81			1981-91		
		Population 1971	Population 1981	Decadal Growth	Population 1981	Population 1991	Decadal Growth
1	2	3	4	5	6	7	8
1.	Olwara (Sawai Madhopur)	953	1193	25.18	1193	1207	1.17
2.	Endwa	1192	1406	17.95	1406	1728	22.90
3.	Shyampura	2541	3066	20.66	3066	3810	24.26
4.	Parli	361	504	49.61	504	724	43.65
5.	Bhadlao	363	551	51.79	557	708	28.49
6.	Kundera	2495	3022	21.12	3022	3544	17.27
7.	Rawal	1286	1596	24.10	1596	2025	26.87
8.	Uliyana	638	803	25.86	803	1070	33.25
9.	Khandoj	43	115	167.44	115	113	15.65
10.	Sherpur	364	548	49.72	548	921	68.06
11.	Kutalpura Maliyan	297	393	32.32	393	524	35.33
12.	Khawa	249	291	16.86	291	429	47.42
13.	Jamool Khera	508	626	23.22	626	822	31.30
14.	Khilchipur	3176	4123	29.86	4123	5019	21.73
15.	Ramsinghpura	865	1227	41.84	1227	1555	26.73

Contd.

TABLE 3.4 : *Contd.*

1	2	3	4	5	6	7	8
16.	Khatipura	835	1103	32.09	1103	1576	42.88
17.	Borif	569	698	22.67	698	1014	45.27
18.	Madhosinghpura	239	342	43.09	342	445	30.11
19.	Nimli Khurd	411	575	39.90	575	762	32.52
20.	Nimli Kalan	419	653	55.64	653	903	38.28
21.	Kalibhat	232	95	-59.05	95	83	-12.63
22.	Kushalipura	116	204	75.86	204	238	16.66
23.	Hajjam Kheri	67	246	267.16	246	226	-8.13
24.	Halonda	459	623	35.72	623	879	41.09
25.	Mordoongri	151	232	53.64	232	298	28.44
26.	Bodal	236	322	36.44	322	443	37.57
27.	Bheronpura	205	138	-32.68	138	159	15.21
28.	Dumoda	634	1220	92.42	1220	1002	-17.6
29.	Baso Kalan (Khandar)	341	422	23.75	422	548	29.85
30.	Baso Khurd	418	559	33.73	559	596	6.61
31.	Bhoori Pahari	1331	1815	36.36	1815	2304	26.94
32.	Dungri	1339	1190	-11.12	1190	1515	27.31
33.	Sawtan	868	734	-14.45	734	940	28.06
34.	Kathooli Dhanapala	174	200	14.94	200	257	29.50

Contd.

TABLE 3.4 : *Contd.*

1	2	3	4	5	6	7	8
35.	Padra Nizamat	!61	204	26.70	204	229	12.25
36.	Goth Bihari	467	635	35.97	635	872	37.32
37.	Sanbar	73	253	246.57	253	314	24.11
38.	Khandar	4324	5856	35.43	5856	7361	25.70
39.	Rawrah	314	425	35.35	425	578	36.00
40.	Jaisinghpura	434	537	23.73	537	687	27.93
41.	Mei Kalan	958	1086	13.36	1086	1365	25.69
42.	Indala	32	46	43.75	46	424	821.73
43.	Jetpur	407	700	71.99	700	1049	49.85
44.	Chhan	2098	3596	71.40	3596	4502	25.19
45.	Allapur	819	948	15.75	948	1202	26.79
46.	Bahrawada Khurd	2246	3025	34.68	3025	3929	29.88
47.	Niwandi (S.M.)	308	510	65.6	510	658	29.0
48.	Bhundeda	491	621	26.5	621	843	35.7
49.	Jeenapur	397	497	25.2	497	681	37.0
50.	Himmatpur	109	166	52.3	166	197	18.7
51.	Qila Ranthambhore	43	68	58.1	68	75	10.3
52.	Sakhwas	249	276	10.8	276	494	79.0
53.	Gopalpura	—	—	—	—	236	NU
54.	Kailashpuri (S.M.)	—	—	—	—	665	NU

villages of the III category (high population and high growth rate). In 1991 their number was 19, with Indala (42) topping the list showing a population of 424 and a growth rate of 821.73%, followed by Sakhwas (52) having a population of 494 and a growth rate of 79.0% (Fig. 3.10).

The changing scenario of population growth in the fringe areas and buffer zone of the sanctuary distinctly shows that drastic changes have taken place in the last two decades and the population movements have been of highest order. Summarily, the causes of these changes have been found to be as given below :

(i) Evacuation of villages of the core areas and their first spread of the buffer/fringe zone of the sanctuary.

(ii) Migration of the initial settlers to better sites at a later stage, which was for better agricultural land, opportunities for self-employment, engagement in secondary and tertiary occupations and proximity to city, roads and railroads, affinities for relatives and people of same caste/clan.

(iii) Villages having less population and less growth rate have not shown marked changes, as they had sufficient land for substantive agriculture.

(iv) Villages with low population but having expanding means of job opportunities and self-employment got the highest number of immigrants and thus their population swelled disproportionately.

Tourism : Growth and Seasonal Rhythms

The Ranthambhore National Park, having the tiger reserve and an operational 'Tiger Project' for the last 25 years (1973-1998), is one of the best tourist sites of Rajasthan where the people can freely see the animal behaviour of tigers and herbivores. The annual record of the Government kept up-to-date by the sale of entrance tickets, shows that during the last 11 years (1987-88 to 1997-98), the number of Indian visitors has risen from 11,136 to 18,646 marking an increase of over 67% (Fig. 3.11). The rise in the number of foreign tourists has been even more phenomenal. From a meagre 3901 tourists in 1987-88, their number in the preceding year 1997-98 has been 17,433 showing an increase of about 367% (Table 3.5).

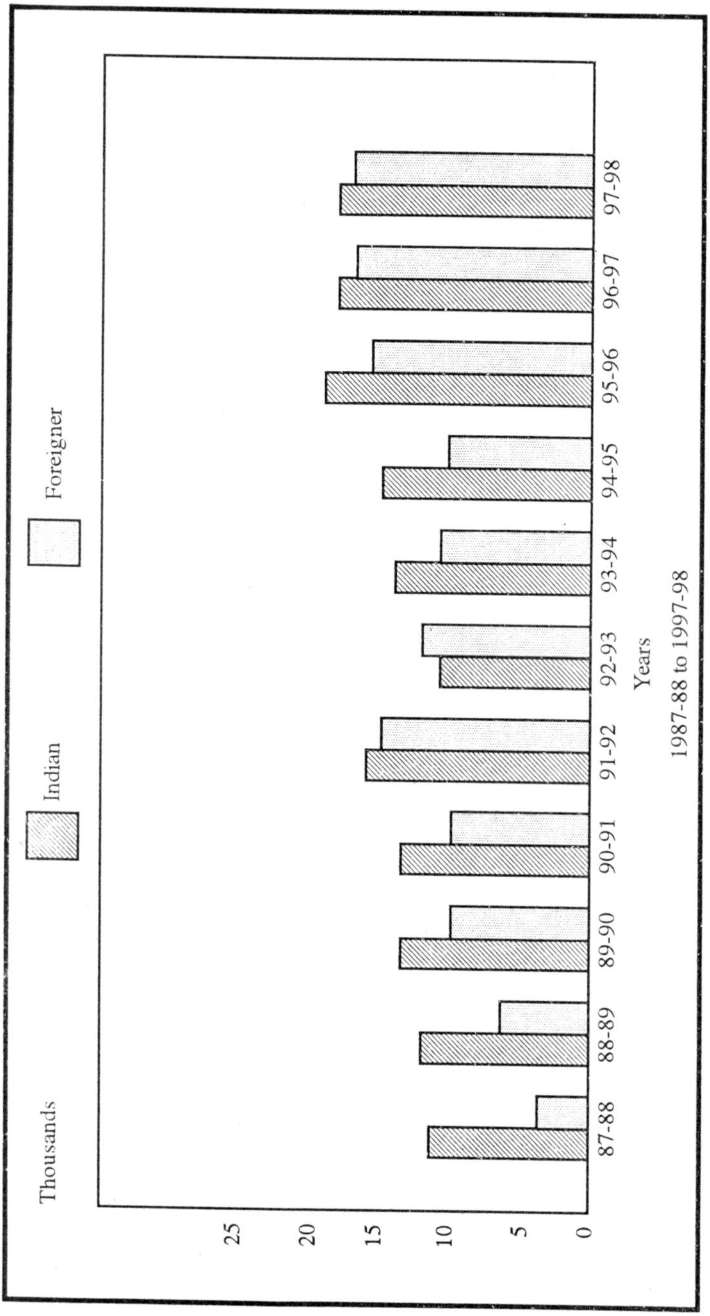

FIG. 3.11 : *Ranthambhore Sanctuary — Annual Tourism Pressure*

TABLE 3.5 : **Annual Tourism Pressure**

Year	Indian	Foreigner
1987-88	11136	3901
1988-89	11880	6474
1989-90	13592	10034
1990-91	13842	10266
1991-92	16302	15284
1992-93	10869	11851
1993-94	13925	10829
1994-95	15630	10534
1995-96	19559	16375
1996-97	18646	17433
1997-98	18646	17433

Source : Department of Tourism, Government of Rajasthan.

TABLE 3.5 (A) : **Growth of Tourists**

Percentage of Growth

1993-94 to 1994-95	1994-95 to 1995-96	1995-96 to 1996-97	1996-97 to 1997-98	Average Growth
68.48	15.26	15.91	43.29	35.73
450.71	-71.93	164.39	-38.29	126.22
46.94	52.72	17.82	6.84	31.08
—	—	—	—	--
—	—	—	—	—
—	—	—	—	—
-5.29	28.12	-5.83	15.76	8.19
27.76	26.12	-5.31	16.48	16.20
9.56	13.68	30.31	-3.70	12.40
37.57	115.51	-22.06	30.64	40.41
21.20	73.15	-7.51	27.54	28.59
34.81	8.45	21.06	-	21.44

It is obvious that such a significant rise in tourism has been due to increasing fame and splendour of this sanctuary, which is noticeable today through increase in tiger population and the maintenance of the sanctuary as per the needs of animal populations of all types.

The impact of tourism is most profound on the animal behaviours as it encroaches upon their tranquillity, disturbs their habitat and their movements from one part to the other in search of food and water.

Although the park authorities have made specified trails which can be transected by visitors and that too in the afternoon, yet the observations show that with the constant movements of jeeps full of enthusiastic visitors are a common and most frequent source of disturbançe to the park animals, which are either is siesta under the tree shades or gathered around the water holes.

A glance at Fig. 3.11 shows that in the last three years (1995-96 to 1997-98) the tourist pressure — both of Indians and foreigners has been stable, fluctuating around 36 thousand or more giving an average of 100 visitors per day which is very high. The Park is closed for public entrance in the rainy season (July through September) for about 100 days due to muddy roads and blockage of certain sites on account of flowing water springs and streamlets. Discounting this period of tourist prohibition, the tourist pressure in the remaining 9 months is substantially high and exceeds 135 visitors per day. However, this number rises much more on Sundays and other holidays, when the people swarm in large nunbers.

The tourists besides disturbing the harmony of wildlife in natural habitats are also a source of spreading garbage and other human used materials like peels of fruits, egg-shells, eatables, cigarette tips, used papers etc. which are likely to be eaten by wild animals, ultimately causing a variety of diseases. The sudden death of over 500 ungulates in 1984 by the outbreak of some unknown disease is yet a constant reminder to prohibit the spilling of such articles and to keep the park under most hygienic conditions in natural ecosystem. Constant interferences by visitors are highly disturbing to the health and tranquillity of animals. When the human induced activities of the villages which were formerly inside the park have been nearly banned, there is no reason to allow the visitors to approach the natural abodes of these animals, if the park has to be maintained in natural splendour.

Seasonal Variation of Tourism

The data taken for 4-year duration (1993-94 to 1997-98) show that a total of 153,827 tourists visited the park in these years giving an average of 38,457 tourists per year.

Among the total tourists a large number 77,062 (47%) visited in the winter season (Fig. 3.12). The time when temperature is low, the weather mild cool, devoid of any rainfall and the park is lush green. Although in the winter season, the tigers rarely come out of their den, except for

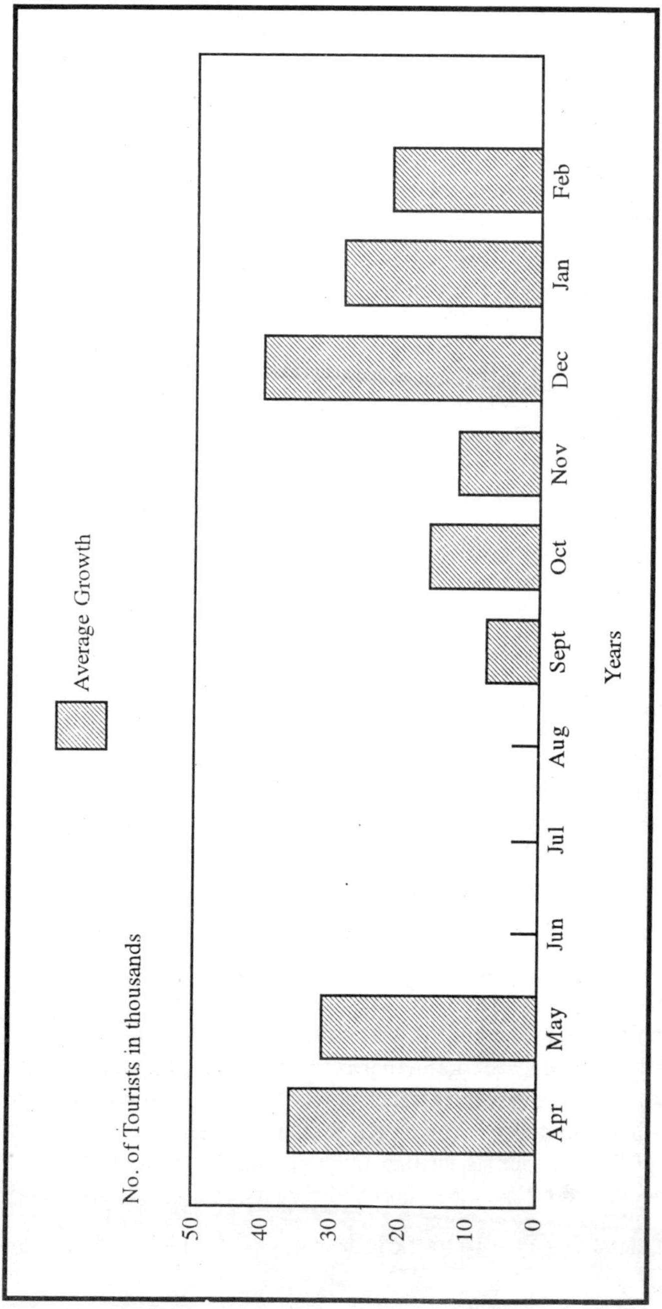

FIG. 3.12 : *Ranthambhore Sanctuary—Average Growth Rate 1993-94 to 1997-98*

occasional basking in the sun or to have a kill to satisfy their hunger. But the herbivores like Chinkara, Sambar, antelopes and deer have their full play near the water holes and the open squares to shun the cold wintry winds.

The spring season lasting from mid-February to the end of March is the second most preferred season of the tourists, when about 38,769 tourists (25.0%) visit the Park. It is the season when the life rejuvenates after the period of winter hibernation. Pairing of Chinkaras and antelopes starts with frequent calls for mating. The grasses are although pale yet they provide enough fodder for them.

Summer is the best season for wildlife to observe when the animals collect near water holes at dawn and dusk providing a rich ecstasy of a rare view. The tigers also come out in the open and reach the water holes to catch their prey. In summers, the number of tourists is 19,409 (7%). The number is lower due to the scorching heat which the visitors avoid.

The number of tourists in autumn is 18587 (21.0%). After the rains, the park roads are muddy, often covered with wild vegetation and the visitors do not get full view of the wildlife. So the number of tourists decreases . However, the flow of tourists starts again in September soon after the rains, when the vegetation rejuvenates and the visitors start coming to watch the natural beauty of this sanctuary (Fig. 3.13).

One of the most obvious reasons of inland tourist influx in the sanctuary in the month of September is the congregation of pilgrims in the Ganesh Chaturthi Mela, which is held in the Ranthambhore Fort every year. Quite a large number of pilgrims avail this opportunity to visit the fort and the sanctuary together.

The monthly trend of tourism for the year 1997-98 in this sanctuary also shows that the number of foreign tourists is larger than the inland tourists in two months viz. November and February (Fig. 3.14). This is the beginning and the end of the winter season in Rajasthan, when most of the foreign tourists visit northern parts of India, specially Agra, Delhi, Bharatpur, Jaipur and Jaisalmer and those who have fascination for wildlife visit Ranthambhore also. This is the reason that one finds the largest number of foreign tourists visiting this tiger sanctuary throughout the winter months from November through February.

Collateral to the tourist traffic is the number of vehicles which enter

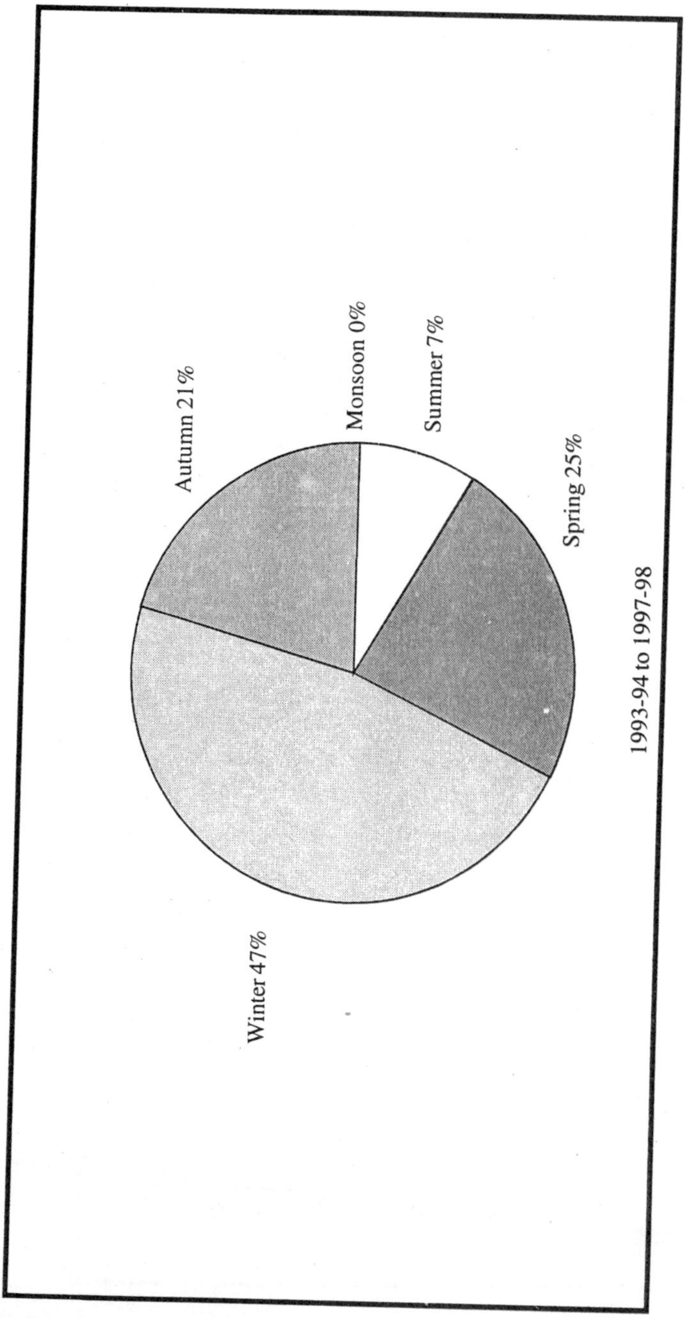

FIG. 3.13 : *Ranthambhore Sanctuary — Seasonal Variation of Tourism*

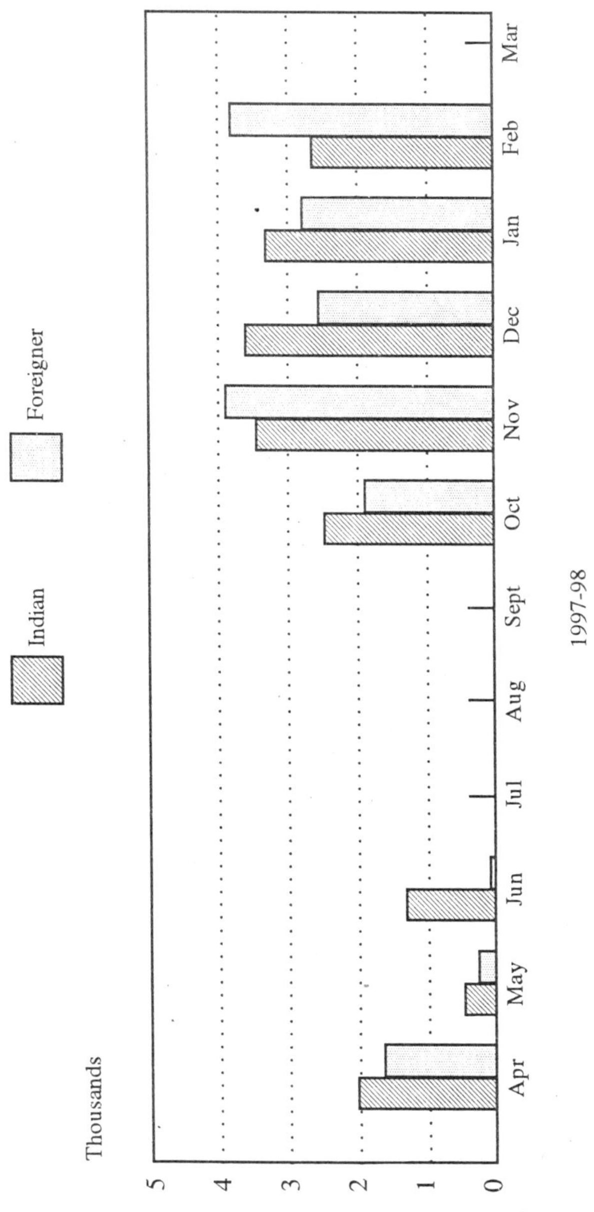

FIG. 3.14 : *Ranthambhore Sanctuary — Monthly Trend of Tourism*

the sanctuary every month. The data of the last 5 years (1993-94 to 1997-98) show that throughout the winter season the number of vehicles plying in this sanctuary every month varies from 820 to 850 (Fig. 3.15). Their number has been constantly increasing from 1993-94 onwards. The highest number of vehicles in November, December and March was in the year 1996-97 which corresponded to that in January, February and April in the year 1997-98. The galloping rate of increase in vehicles is attributable to the growing number of visitors. It can hardly be denied that these vehicles are a cause of disharmony to the wildlife and that their behaviour invariably changes with these interferences.

Tourism Dynamics and Future Trends

Tourist traffic is maximum in winter months and continues to decline each month after the spring season. This monthly rise and fall taken for the last 5 years (1993-94 to 1997-98) depicts a dynamic picture especially for winter months (October to March), the period when there has been a phenomenal rise (Fig. 3.16). In October, the number of visitors increased from 3304 in 1993-94 to 4370 in 1997-98 showing a 5-year increase of more than 32%. In this month, there has been a slight fall in 1994-95 and 1996-97 which is only marginal (Table 3.6).

In November, December and January months, the rise during this 5-year period has almost been constant recording an increase of 78%, 56% and 201% respectively. The momentum of tourist traffic continues in February and March also, although with lesser number. Nevertheless the rise in February for the aforesaid 5-year period is 148% and that in March is 77%.

It is obvious therefore that bulk of the tourists now come from October through March and have a peak season in November (7369 tourists in 1997-98), which gradually declines in December (6131), and January (6122) but rises again in February (6442). Mild winters of November and February are thus the most favourable months to visit the Ranthambhore Sanctuary and so preparation should be made in advance for their accommodation, hospitality, provision of vehicles and also to make itinerary of their visit to various sights of the sanctuary. At present the vehicles plying inside the sanctuary do not follow the route allotted to them and take to the jungle trails in a zig-zag manner in their effort to see a tiger and thus disturb the harmony of wildlife.

The quantum leap in tourist numbers has inevitably been on account

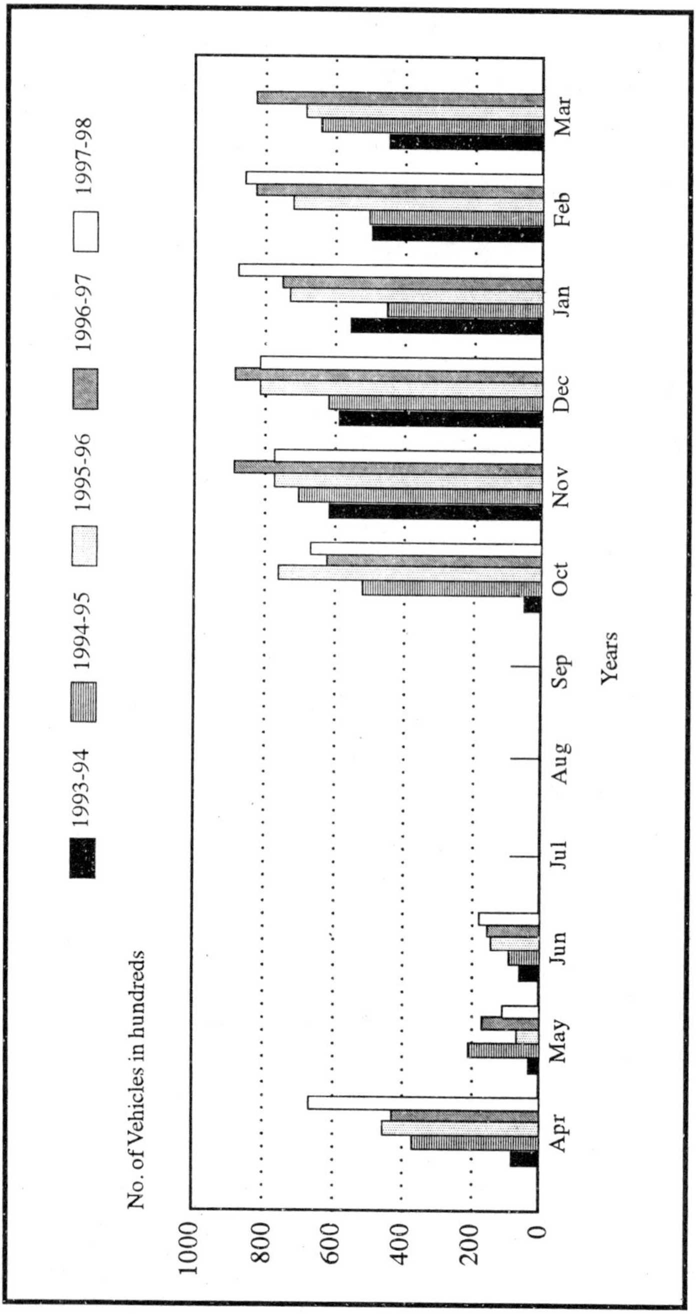

FIG. 3.15 : *Ranthambhore Sanctuary — No. of Vehicles*

TABLE 3.6 : **Tourism Dynamics**

Month	1993-94			1994-95			1995-96			1996-97			1997-98		
	Indian	Foreigner	Total	Indian	Foreigner	Total	Indian	Foreigner	Total	Indian	Foreigner	Total	Indian	Foreigner	Total
April	568	571	1139	1059	860	1919	1251	961	2212	1249	1315	2564	1998	1676	3674
May	175	34	209	906	245	1151	286	37	323	694	160	854	381	146	527
June	448	27	475	637	61	698	948	118	1066	1221	35	1256	1304	38	1342
July	—	—	—	—	—	—	—	—	—	—	—	—	—	—	—
August	—	—	—	—	—	—	—	—	—	—	—	—	—	—	—
September	—	—	—	—	—	—	—	—	—	5	44	49	—	—	—
October	1685	1619	3304	1734	1395	3129	2519	1490	4009	1789	1986	3775	2436	1934	4370
November	2657	1529	4146	3261	2036	5297	3191	3496	6681	3262	3064	6326	3472	3897	7369
December	2435	1487	3922	2897	1401	4298	2973	1913	4886	3756	2611	6367	3647	2484	6131
January	1088	940	2028	1601	1189	2790	3632	2381	6013	2341	2345	4686	3315	2807	6122
February	1053	1541	2594	1450	1694	3144	2332	3112	5444	2052	2983	5035	2612	3830	6442
March	1264	1623	2887	2189	1703	3892	1748	2473	4221	2555	2555	5110	DNA	DNA	DNA

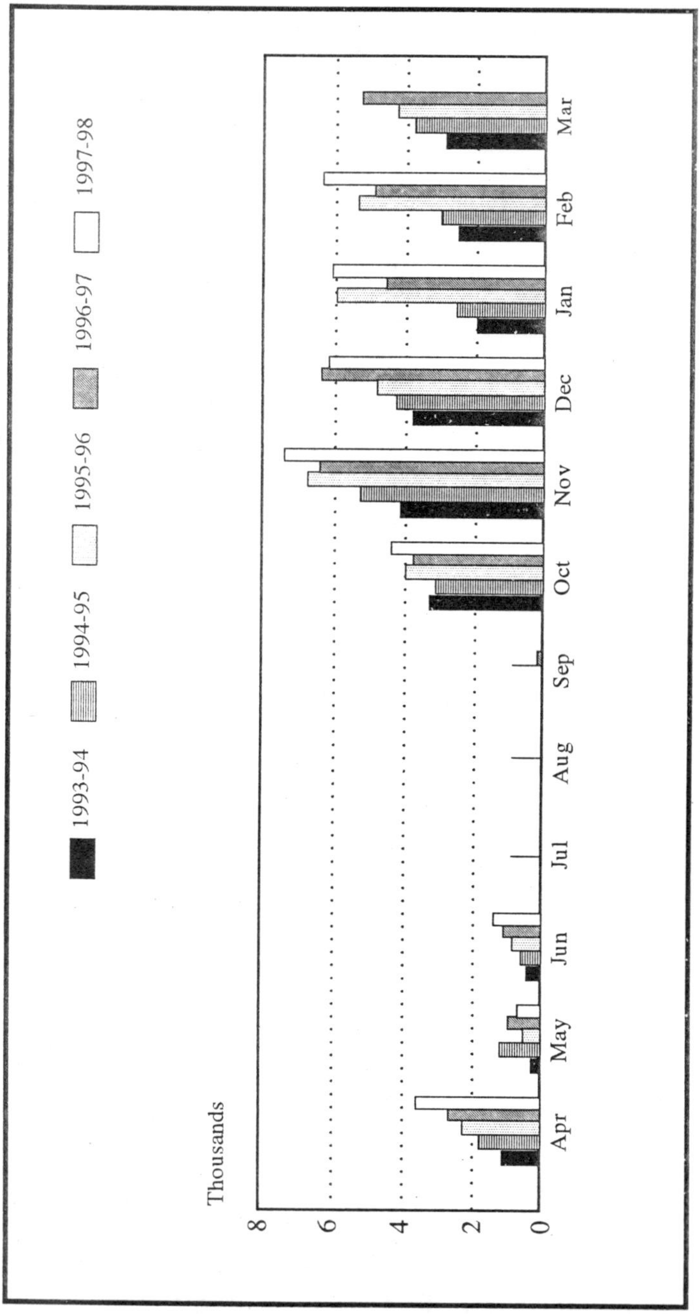

FIG. 3.16 : *Ranthambhore Sanctuary — Tourism Dynamics*

of the unprecedented rise in the number of foreign tourists. In November 1997, the number of Indian and foreign tourists was 3472 and 3897 respectively and the foreign tourists outstripped the Indians by more than 12%. In December and January months, the Indian tourists are more than the foreigners. Their number in December 1997 was 3647 and 2487 respectively while in January 1998, the number of Indian and foreign tourists is 3315 and 2807 respectively. But again in February 1998, the number of Indian and foreign tourists was 2612 and 3830 respectively showing a spurt in the number of the latter.

It is also true that the foreign tourists are likely to increase further, mainly in the mild winter months, when it is the time for them to come to India and to enjoy leisure in the pre- and post-Christmas vacations and to have the warmth of the day when their own countries are freezing with cold.

To increase the foreign tourist visits within the controlled system of ecology and without impinging the natural harmony of wildlife is a moot point of consideration to plan for future. Some strong positive efforts in this direction have to be made, which may also need institutional changes and drastic deviations from the past practices and conventions. Some of the salient aspects of these changes may include (a) rise in the entrance fee both for the foreign and Indian tourists, (b) allocation of definite jungle trails to the jeeps plying inside the park, which are to be strictly followed, (c) permitting only a few vehicles at one time and for a limited period (1 hour), (d) use of electric-driven silent vehicles to reduce the stormy noise of diesel-driven jeeps, (e) stopping all edibles, cigarettes, water bottles, leather articles etc. at the gate and screening the entrants for this purpose, (f) starting air-balloon flights of half-an-hour duration to have an overview of the wildlife without disturbing them on the ground.

The Vehicular Traffic

It is not the tourists alone but the vehicular traffic of the gypsies and canter autovans that disturb the tranquillity of this sanctuary. These vehicles speeding up their jungle trails to catch the early site of tigers and other animals move about in an irregular way often negotiating the precipitous slopes and create unbearable noise of the vehicular engines. The data of 5-year-period (1993-94 to 1997-98) in this regard show that there has been almost a double increase in their numbers (Fig. 3.15).

The total vehicles that entered the sanctuary in 1993-94 for the 9

TABLE 3.7 : **Annual Trend of Vehicles**

Month	1993-94			1994-95			1995-96			1996-97			1997-98			Total			
	Canter	Gypsy	Total	Canter	Gypsy	Total	Canter	Gypsy	Total	Canter	Gypsy	Total	Canter	Gypsy	Total	Canter	Gypsy	Total	Average
April	31	47	78	49	327	376	49	401	450	51	377	428	105	557	662	285	1709	1994	389
May	21	4	25	25	181	206	25	39	64	38	134	172	18	92	110	127	450	577	115
June	43	2	45	23	62	85	26	116	142	35	113	148	52	132	184	179	425	604	121
July	—	—	—	—	—	—	—	—	—	—	—	—	—	—	—	—	—	—	—
August	—	—	—	—	—	—	—	—	—	—	—	—	—	—	—	—	—	—	—
September	—	—	—	—	—	—	—	—	—	—	—	—	—	—	—	—	—	—	—
October	77	389	46	74	443	517	183	590	773	114	505	619	150	528	678	598	2455	3053	611
November	120	494	614	124	603	727	150	636	786	197	694	891	260	528	788	851	2955	3806	761
December	104	492	596	101	518	619	195	627	822	184	708	892	184	633	817	768	2978	3746	749
January	63	496	559	82	370	452	139	602	741	136	623	759	197	695	892	617	2786	3403	681
February	65	434	499	91	410	501	142	588	730	136	694	830	212	662	874	646	2788	3434	687
March	61	488	549	113	525	638	105	582	687	140	685	825	DNA	DNA	DNA	559	2280	2839	568
Total			3011			4121			5195			5564			5005			22896	
Average (Monthly) (For 9 months)			335			458			577			618			625			509	509

months when it is open was 3011 while that in 1996-97 it was 5564 marking an absolute increase of 2553 (84.8%). The data in this regard for the month of March 1998 are not available and so the fiscal year 1997-98 has not been taken into account. But nonetheless, it is assumed that the absolute numbers for 1997-98 will be higher than the preceding year.

A further analysis of the data shows that the average monthly number of vehicles per year is increasing every year (Table 3.7). It was 335 in 1993-94 which rose to 625 in 1997-98 marking an increase of 86.6% in the past 5 years. It may well be acclaimed that this monthly average is likely to cross the 800 mark in the year 2000. On the basis of the existing data it may be asserted that about 21 vehicles enter the reserved park every day and that too between 2 to 6 p.m., making a hub of vehicular traffic inside and creating thunderous noises all round.

The monthly averages of the data of vehicular traffic further show their highest number in November (761) followed by December (749), February (687) and January (681). Such a data reveals that in November the average daily vehicular traffic is more than 25 and that in the other winter months it hovers around 24 or so.

Needless to mention here that these vehicles besides creating noise, also spread the air pollution by burning the fossil fuels and very often their leakages of diesel, petrol and fuel oil on the ground become a potent danger to wildlife.

The tourist traffic is likely to increase in future and so also the vehicular traffic. While it may not be possible to shut the doors of the sanctuary for the welfare of the wildlife it should be possible that electric vans which are noiseless and pollution-free, cable cars in between mountain ranges and balloon flights may be arranged for the visitors as an alternative to the diesel-driven vans. It would also be more safe to the visitors also to save themselves from any possible attack of wild animals which at times are in privacy and siesta.

4

DISCUSSION AND ANALYSIS

Environmental Impact Assessment

AN integral component of the present study is to analyse the impact of the population upon the sanctuary conservation both in terms of their own perceptions as well as the observations made in the present survey. For this purpose seven villages were selected for field observations and to explore the perceptions of the local inhabitants in regard to their needs of fodder and fuelwood and also for the conservation of this sanctuary. It has to be emphasized time and again that conservation of Ranthambhore sanctuary is of utmost priority in Rajasthan, where more than two-third area is desert and semi-desert and the constant threat of increasing desertification by human ambivalence is ever widening. More anxiety is now expressed to save the fringe areas of the Thar desert from ecological degeneration and population impingement to arrest the sprawl of the desert and intensification of activities promoting the process of desertification. With the meagre remains of natural forests and primeval vegetation in Rajasthan, there is a dire necessity to conserve these remaining isolated patches of natural vegetation to save the population from the consequences of ecological disequilibrium.

The seven villages selected for the evaluation of the impact of vacating the sanctuary precincts and the methods adopted to conserve and regenerate the Ranthambhore sanctuary are Indala, Mordoongri, Bodal, Katholi, Padra, Gopalpura and Kailashpuri, out of which the first-five are still inside the sanctuary and the latter two are the new villages established by the government to resettle the evacuated population of the 12 villages formerly inside this sanctuary. It may be mentioned that Indala, Katholi and Padra villages exist even today in the core area, Mordoongri lies in the buffer zone, Bodal and Gopalpura both touch the boundary of the sanctuary and Kailashpuri is located at quite a distance from the boundary.

Profiles of Sample Villages

1. *Indala*

Gurjar-dominated Indala village is located on the high plateau of Upper Bhander Sandstone in the southern part of the Ranthambhore Sanctuary. The village is approachable by jeep only through core area, covering a distance of about 30 kilometres. The southern face of the plateau is 300 metres high from plain area and is hardly 2 kilometres from Chhan village, a village situated at the foothills of Sandstone Plateau. The population as per 1991 census was 424 consisting of 73 households, but the present survey reveals that there are only 71 persons living in 16 households. The total area of the village is 105.01 hectares. The occupation of villagers is agriculture and animal husbandry. The soil is thin having reddish black colour mixed with stones. The land is fertile but for lack of irrigation facilities only rainfed crops are grown. There is a heavy pressure on forest resources of the core area on account of the fuelwood demand, grazing and expansion of agricultural land on forest land within the periphery of 4-5 kilometres (Plates).

The village is under process of shifting to other site out of the Sanctuary. The survey reveals that majority of people are ready to leave this place in view of dark future provided government fulfil their demand for legitimate compensation.

2. *Mordoongri*

Mordoongri is also in the southern part of Ranthambhore Sanctuary. This village is situated near Sawai Madhopur-Khandar road. It is a valley-sited village having rich agricultural land and good irrigation facilities. This village has been thrice displaced for construction of Mansagar and now it is again in the process of evacuation from the core area under forest policy.

The population as per 1991 census was 283 having 63 households. The village area is 548.76 hectares.

The water table is very high and there are 5-6 tubewells to irrigate the fields. Apart from the availability of electricity and metalled road, the village faces the problem of destruction of crops by wild animals. Sometimes leopards take up their goats and calves. It has been reported that there is no future of agriculture in this village. Gurjars are dominating the population followed by Nayaks and Keers. The village has a primary school under "Shiksha Karmi Yojna" with 2 teachers and 57 students on roll. There are 16 biogas plants but most of them are out of service.

3. *Katholi*

Katholi is accessible by non-metalled circuitous route encircling the Ranthambhore Sanctuary from north connecting Doongri and Basu. The village is situated in the core area on hill top. The area of the village is 827.95 hectares. The population as per 1991 census was 259 consisting of 49 households but as per our survey (1997) it is 300 having 60 households. 90% population is of Gurjars and the remaining are Kumhars, Brahmins and blacksmiths.

Agriculture is the main occupation. Land is fertile. Water is abundant in 8 wells. Pumpsets are 15. The cattle population is about 1130. Lifting of cattle by wild animals is the biggest problem of this village. Encroachment on forest land is common.

4. *Padra*

It is two kilometres south of Katholi village. The terrain is undulating. Hogback ridges are common. The area is 1154.80 hectares. The population is 300 consisting of 60 households. This village is also Gurjar dominated. Agriculture is the main occupation. Villages are dependent for fuelwood, grazing and other demands on forest resources. 8 wells and 10 pumpsets are available for irrigation. Cattle population is 500.

5. *Gopalpura*

Gopalpura is a newly established village of Khandar tehsil near Pharia road. In 1979 four villages Nagdi, Lahpur, Chanden-Chindeli and Kila Ranthambhore were shifted from core area and resettled at this site. The area of the village is 951.97 hectares having 236 population and 57 households. The land of the village being stony, degraded without vegetation and poor agricultural land does not bring back the dignity and prosperity of villagers.

6. *Kailashpuri*

Kailashpuri is also a resettled village far from the core area nearly 6-7 kilometres south of Sawai Madhopur-Chhan road. The area of the village is 1229.66 hectares with 665 population and 110 households. It is a well-planned village on Piedmont zone of the Aravalli hills. The government also provided facilities as they had in the core area. The agricultural land is fertile. Water is sufficient for their domestic and other requirements. The resettlement has given them an opportunity for exposure and cross culture development. The village also has the problem of lifting of cattle and destruction of crops by wild animals. However, the villagers enjoy better facilities than Gopalpura.

7. *Bodal*

Bodal is situated at the periphery of Ranthambhore Sanctuary on Sawai Madhopur-Khandar metalled road. The total area of the village is 1706.67 hectares of which 1541.15 hectares is under forest cover and only 114.52 hectares is available for cultivation. Population as per census 1991 was 443 with 74 households. The agricultural land is fertile but there is a constant threat of wild animals to livestock and crops. Some local inhabitants are engaged in stone quarries in spite of total ban on mining activities in the Sanctuary.

Population Pressure

The 1991 population of all villages varies from 259 (Katholi) to 665 (Gopalpura). At the time of field survey for the present study, the population was found to be highly variant in all the villages. Thus Indala had only 71 persons (424 in 1991) and Mordoongri 283 (298 in 1991) showing thereby that in the last 5 years more persons have emigrated from them. However, the population of Bodal (512), Katholi (295), Padra (304), Gopalpura (269) and Kailashpuri (665) has increased in the corresponding period. The population growth in the last 16 years (1981-97) has been exceptionally high in all these villages ranging from 22.0% (Katholi) to 59.0% (Bodal) showing that these village sites are still the favourite spots which attract the migrants to settle (Table 4.1).

Agricultural Land

All the above seven villages have fairly good areas of agricultural land to provide ample opportunities to inhabitants to do substantive farming. While Padra (120 hectares) has the lowest agriculture land, Kailashpuri

TABLE 4.1 : **Sample Villages : Impact Assessment**

Specifications		Exists in Core Area	Exists in Buffer Zone	Touches Park Boundary	Exists in Core Area	Exists in Core Area	Shifted Village touches Park Boundary	Shifted Village at quite distance to park territory
Village Surveyed		Indala	Mordoongri	Bodal	Katholi	Padra	Gopalpura	Kailashpuri
1		2	3	4	5	6	7	8
Population Pressure	Population 1981	46	,232	322	200	204	—	—
	Population 1991	424	298	443	259	236	665	—
	Decadal Growth 81-91	821.7%	- 28.4%	37.6%	29.5%	12.3%	—	—
	Population 1997-98 (at the time of survey)	71	283	512	295	304	269	665
	Growth in 16 years 81-97	54.3%	22.0%	59.0%	47.5%	49.0%	14.0%	19.2%
	Total No. of Household at survey time - 1997	16	63	67	56	60	57	125
Agriculture Area in hectares inverse to Natural Ecosystem		125	214.00	114.52	160 encroaching on forest land	120	300.50	370.66
Pressure on water Resources — means used	Irrigated Land	—	224.05	285	618	125		
	Hand-pumps	—	✓ 2	✓ 2 working & dry each	-2 dry	2 working	✓ 1 working 2 dry	—
	Wells : Potable use / Irrigation use	—	5 + 6 = 11	2 + 10 = 12	1 + 7 = 8	1 + 8 = 9	1 + 40 = 41	1 with pump set
	Reservoir/Pond	2	1	1	—	—	1	1
	Tankers	1	—	—	—	—	—	—

Contd

TABLE 4.1 : ***Contd.***

1		2	3	4	5	6	7	8
Grazing pressures Field survey - 1997	Bovines	122	347	368	310	426	390	1018
	Sheep & Goat	315	219	246	793	378	328	920
Human Cattle Ratio		6.15	2.0	1.2	3.7			
Occupation of households Characterise activity of villages concerning with forests (Based on Number of households)	Cultivator	40	58	49	59	55	48	50
	Agriculture Labour	08	10	09	09	14	12	10
	Cattle grazing	12	10	08	12	16	15	14
	Dairy	20	15	19	15	14	10	12
	Petty Shopkeeper	—		06	—	—		
	Other If yes specify	—	old history writing	09	05	01	15	14
Fuelwood Pressure (Estimated)	Total Yearly Consumption	345 (Quintals)	664.4 (Quintals)	2088 (Quintals)	1439 (Quintals)	1575 (Quintals)	1470 (Quintals)	1575 (Quintals)
	Per Capita Yearly Consumption	485.9 kg.	234.8 kg.	407.81 kg.	487.7 kg.	518.2 kg.		
Number of trees cut down/lopped in a year approximately	Total trees	390	643	607	2183	2637	2120	2150
	Per Head trees	5.45	2.27	1.19	7.4	8.7	3.1	2.6

Contd.

TABLE 4.1 : ***Contd.***

1		2	3	4	5	6	7	8
Awareness to minimise Fuel consumption	Existence of awareness	No	Yes	Yes	No	No	Yes	Yes
	knowledge of substitutes	—	Biogas Solar Energy	Biogas	—	—	Biogas	Biogas
Use of Chemical Fertilizers (Ecological Effects)	Urea (in kg.)	—	5325	12000	26000	9375	15200	16500
	Diammonium (in kg.)	600	2885	5375	12715	4210	10320	11200
General View of Family Planning		Positive	+ive (73.0%) 11.3% No response	+ive (86.56%)	+ive (56.5%)	+ive (53.8%)	+ive (74.2%)	+ive (88.4%)
Environmental Awareness		Generally found	Generally found	Generally found	Generally found	Generally found	Generally found	Generally found
Eco-destructive activity around the village	Industrial	—	—	—	—	—	Brick kiln	—
	Mining	—	—	Stone Quarries (Unorganised)	—	—	—	—
	Other	—	—	—	—	—	—	—

Contd.

TABLE 4.1 : ***Contd.***

1		2	3	4	5	6	7	8
Pressure on Park (as surveyors assess with consulting various reliable sources)	Habitation	✓	✓	—	✓	✓	—	—
	Agriculture	✓	✓	—	✓	✓	—	—
	Grazing	✓	✓	✓	✓	✓	✓	—
	Herb collection	✓	✓	✓	✓	✓	✓	—
	Fodder collection	✓	✓	✓	✓	✓	✓	—
	Fuelwood collection	✓	✓	✓	✓	✓	✓	—
Villagers Attitude towards Sanctuary (Surveyors' Assessment)	Improper Entry	✓	✓	✓	✓	✓	✓	—
	Improper Conduct	✓	✓	✓	✓	✓	✓	—
	Hunting	—	—	—	—	—	—	—
	Cause Fire Hazards	✓	✓	—	—	—	—	—
	Use of Explosive	—	—	—	—	—	—	—
	Illegal weapons	—	—	—	—	—	—	—
	Destruction of Habitat	—	✓	—	✓	✓	—	—
	Grazing/entry of cattle	✓	✓	✓	✓	✓	✓	—
Is village an obstacle in free migration of wild animals		No	Yes	Only Agricultural land	Yes	Yes	No	No

(370.66 hectares) has the highest. Since Kailashpuri and Gopalpura are villages of resettled households, it is likely that agriculture land in both of them will increase further but this will be an encroachment upon the forest land or the barren land which will have adverse effect upon ecology. Nonetheless, the existing agricultural land in the villages inside the sanctuary and its persistent use in agriculture is a point of dismay which should be made clear to the native population with respect to ecological benefits. On the one hand, these lands are under constant threats from wildlife who eat away their harvests and on the other they degenerate the ecosystem by spills of chemical fertilizers, pesticides and various types of manures.

Water Resources

Use of water for human and animal drinking purposes, together with irrigation is undoubtedly a pressure on water resources in this area. Installation of hand-pumps, which are 2 each in every village and sinking of open dug wells in large numbers is depleting the water resources at a higher rate beyond the capacity of recharging. The number of wells is exceptionally high in all the villages e.g. Mordoongri (11), Bodal (12), Katholi (8), Padra (9) and Gopalpura (41), which are used both for drinking and irrigation. As the bedrocks of this area are hard and impervious, most of the wells have shallow water levels which soon dry up in summers.

The impact of drafting out the water for human use is most telling upon the natural tanks and reservoirs of the sanctuary which are used by wildlife for drinking. The natural reservoirs like Milak Talab and Galai Sagar become dry in summer season and the forest authorities have to provide water from tubewells to save the wild animals.

In some of the above villages ponds and water reservoirs also exist as in Indala, Mordoongri, Bodal, Gopalpura and Kailashpuri. These are implicitly used by domestic animals for drinking.

Grazing Pressure

According to 1997 survey, the population of cattle, sheep and goats continues to be high (Table 4.1). The human-cattle ratio in the 4 villages inside the park varies from 1.2 (Bodal) to 6.15 (Indala) which is more than the state average (1: 1.2). It shows that the native population is still largely dependent on cattle wealth and that the domestic animals provide substantial surrogatory economy to the people. High population of domestic animals speaks about the higher grazing pressure and also a constant encroachment on natural vegetation.

Fuelwood Requirements

Estimates in this regard show that the yearly fuelwood needs of the surveyed villages vary from 345 qts (Indala) to 1575 qts. (Kailashpuri), the variants depending upon the size of the population, the use of fuelwood and other alternate methods of energy (Table 4.1). All the fuelwood is invariably being cut from the trees of the sanctuary and quite likely some of it is being sold outside also to earn livelihood.

The number of trees cut or lopped every year for fuelwood are exorbitantly large varying from 390 (Indala) to 2637 (Padra) and unless their regeneration is at the same rate, ecological destruction is likely to continue in future. Estimates in this respect show that the per capita highest number of trees cut/lopped is in Padra, possibly for the reason that fuelwood is traded out of this village to the nearby villages and towns just across the river Banas which is nearby. The per capita cutting/lopping of trees in other 6 villages is also high and highly dismal to the ecological restoration of this sanctuary.

Awareness of ecological maintenance in some of these villages has not yet crept in. While the natives of Indala, Katholi and Padra are totally ignorant about ecology and the use of eco-friendly methods to meet with their energy requirements by alternate means, the awareness exists in the rest of the 4 villages, which use biogas and solar energy as a substitute of fuelwood. in Mordoongri solar energy is also used as a means for domestic energy requirements.

Ecological Effects

Harmful effects of chemical fertilizers like urea and diammonium are well known and need not be elaborated here. But special attention has to be paid for their use in this area which is a closed ecosystem, surrounded by hills on all sides with no escape channels. The runoff of the agriculture fields having toxicity of these chemicals and its mixing with the water of natural reservoirs used for drinking by wild animals is an increasingly potent danger and may lead to large-scale extinction of plant species and threat to the animal health. Large doses of such toxic materials may become lethal and cause large-scale mortalities. To avoid all these risks to animals and plants, it is important that these villages be evacuated and the use of chemical fertilizers and pesticides be immediately stopped.

Grazing Pressure

Grazing of domestic animals in the buffer zone of the sanctuary is a common feature, much to the dislike of the sanctuary authorities. In the early eighties, grazing was illegally done in the core area also, which led to scuffles and feuds between the authorities and the villagers, sometimes leading to use of firearms also. Today, grazing is totally banned inside the core area but stray cases of grass cutting and grazing happen daily, either by connivance of the forest guards or by illegal means. The reasons for this illicit grass cutting and grazing are obvious. Firstly, the cattle population in the adjacent villages is very high which cannot be fed by stall feeding alone and secondly, the domestic animals need to go to nearby grasslands for their better health and grazing of green foliage.

Examining the cattle wealth of all the 54 villages it may be said most assiduously that the cattle density is very high in some of the villages. It is more than 300 cattle/km^2 in some villages like Endawa (333), Rawal (314), Khilapa Maliyan (408), Kathipura (413), Dumoda (334), Talwada (319), Allapur (396), Bhundeda (649), Himmatpur (953) and Sakhawas (340). Similarly, a large number of villages lie in the category of having cattle population of 200 to 300/km^2, which too is very high and their needs cannot be met with stall feeding alone.

By far, the highest animal population is of cattle and buffaloes; both are milch animals and so need a standard grass weight of at least 6 to 10 kg/day for daily feed. Nearly half of these animals do not give milk and are therefore left stray to seek their own feeds upon common proper lands, permanent fallow lands and the forest fringe. Cattle lifting by tigers also happens sometimes when the domestic cattle impinge upon the tiger domains.

For the purpose of the present study, grazing pressure has been calculated on the basis of cattle units. One cattle unit consists of one cow/ox/bull or 0.5 buffalo, or 5 sheep/goats or 0.2 camel, which is a standard unit to measure the fodder requirements. The data of cattle wealth are of the year 1988, which were available at the Revenue Board, Ajmer. It may be assumed that in 1998, cattle number have further increased and are in no way less than those 10 years ago.

The highest density is observable in the northwestern parts of Sawai

Madhopur block. It is specially high in the close vicinity of Sawai Madhopur Town, where the supplies of milk and milk products can be profitably marketed and also they can be sent by rail to nearby towns like Kota (Fig. 4.1).

Elsewhere cattle density is higher in the isolated villages located in the south like Kailashpuri, Indala etc. It may also be pointed out that the population of sheep and goats in these villages is also very high. As these two animals can survive even in scarcity areas and can feed upon the small blades of grasses and their roots in the hot summer season, they are very much reared in the Dang areas, which have low agriculture outputs. However, both sheep and goats are heavy grazers to decelerate the further growth of grasses. These two animals make the land barren very soon in which regeneration of grasses takes a longer time.

Cutting and Grazing Pressure

The stress upon the fringe of the sanctuary is a two-pronged problem. Firstly, the grazing pressure of large animal population and secondly, the pressure of cutting and tree lopping by the human population which needs fuelwood and timber.

The survey in this regard shows that the entire sanctuary and the area around it can be divided into 5 classes, which is based upon the vegetative growth and the regeneration of vegetative cover.

(i) *Poor regeneration after 1960* : This is a broad strip which extends from the north of Khandar village up to the banks of Banas river where population and cattle pressure are the highest and regeneration of tree and grass cover has not been satisfactory due to constant cutting of trees and incessant grazing. This type of situation also exists in three villages of the northern boundary and the villages to the south of Sawai Madhopur Town (Fig. 4.2).

(ii) *Wholly or Partly Degraded up to 1980* : This is the belt to the south of Khandar village stretching all along the southeastern fringe of the sanctuary. Isolated patches also exist to the south of Man Sagar and Sawai Madhopur town. These areas are today highly degraded of natural vegetation and further regeneration is not in sight.

(iii) *Affected by Cutting and Grazing* : These areas form a contiguous belt around the core area of the sanctuary. These are parts of buffer

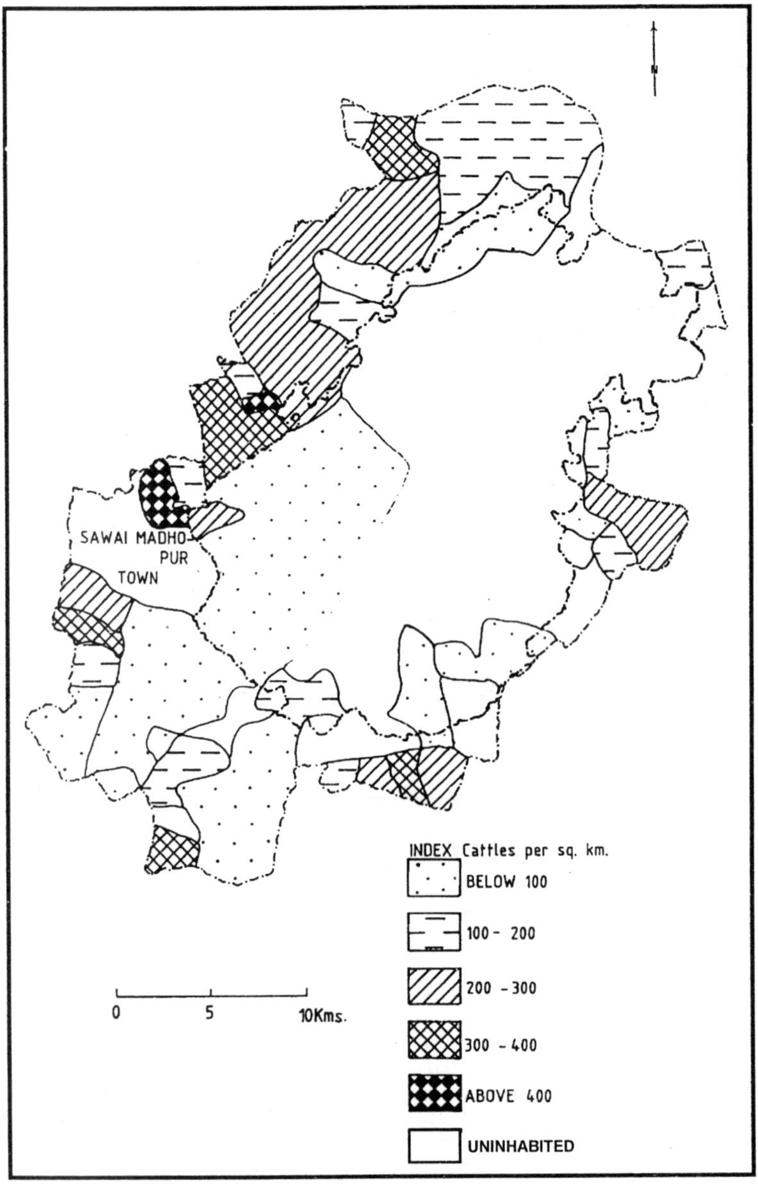

FIG. 4.1 : *Ranthambhore Sanctuary : Grazing Pressure 1988*

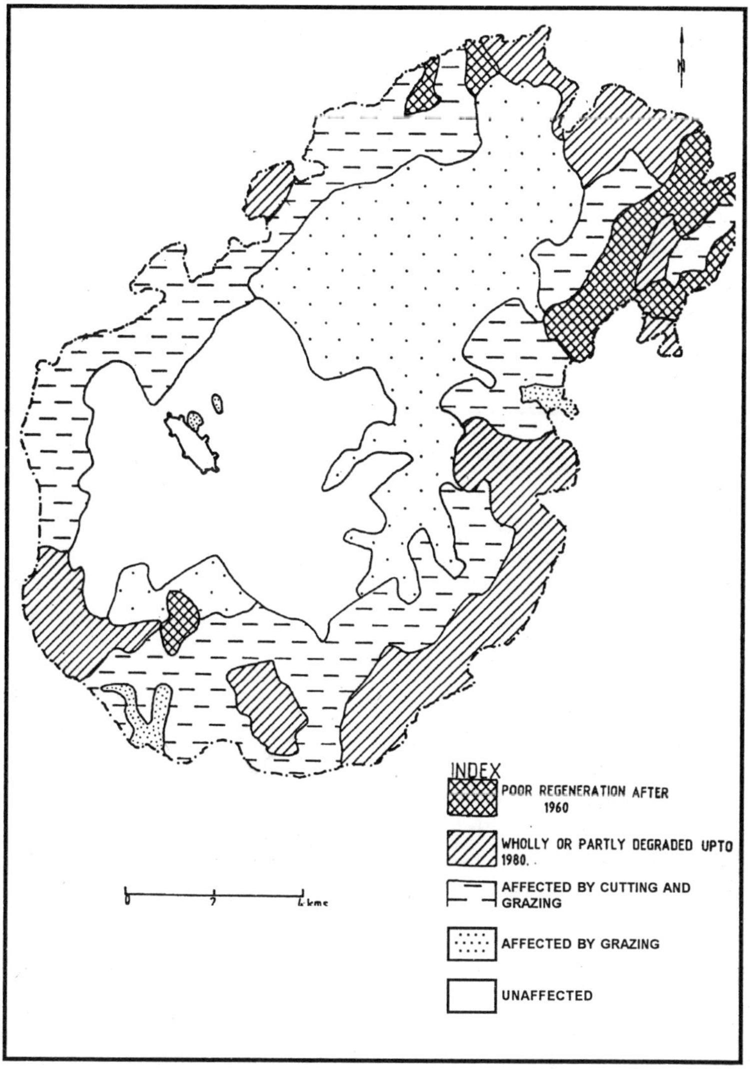

FIG. 4.2 : *Ranthambhore Sanctuary : Cutting and Grazing Pressure*

zones, which have been constantly affected by tree cutting and animal grazing. Due to the poor regeneration of vegetation on the outer fringe, these areas of inner fringe are now under biotic stress and have therefore to be looked after most vigilantly to avoid the further incoming pressure.

(iv) *Areas Affected by Grazing* : These areas exist inside the main sanctuary particularly in the north which has a mountainous terrain that gets grassy cover in the monsoon and the post-monsoon season, due to poor regeneration capacity on the hilly slopes and the pressure of sheep and goats which can easily negotiate these heights, the effects of grazing are markedly noticeable.

(v) *Unaffected Areas* : These areas are in the close vicinity of Ranthambhore fort, Jogi Mahal, Bagola valley, and the compartments of Lakarda, Lahpur, Kachida, Bherda etc. These areas are under the constant vigil of the forest guards and sanctuary authorities and are therefore uninvaded by human impingement. Further, these areas have the largest population of wildlife herbivores and therefore being constantly trekked by visitors in jeeps. So cutting of trees and grazing of grasses is not possible.

Looking into the stress of human and cattle population on one side and the need of conserving the vegetation and wildlife population on the other side, the situation is most dubious. It is yet a dilemma to find the pragmatic means to conserve the sanctuary and also to meet with the needs of the people. Efforts should be made in this direction so that the two opposite systems come to an equilibrium and without affecting the ecology, the economic needs are fulfilled.

Poaching and Illicit Killing of Wild Animals

Wild animals like tigers, antelopes, deer, bear, boars and civets which inhabit this sanctuary have remained under constant threat of illegal killings ever since this sanctuary was formed and even earlier during the reign of former Maharajas and chieftains who used this wildlife reserve for game and shooting. The skin and heads of these animals have been the pride possessions of Maharajas as trophies to speak about their skill in hunting and the fancy for these rare boutiques to decorate their halls of public and private audience. Nevertheless, with the promulgation of Wildlife Act in 1964 and more stringent Act of 1972, today there is a total ban on hunting and game and wildlife is supposed to be absolutely immune from all dangers from 'Man'.

Despite these most stringent Acts, the awareness among the common masses for wildlife conservation and spread of bio-diversity, certain animal species are still under constant danger of poaching either for flesh as the various herbivores or for skin, claws and the concentrated fleshy juices, of tiger for its fur and potency increasing medicines. Further, these articles fetch fabulous prices and are commonly sold in the markets of China, Vietnam, Japan and Southeast Asia. It is not surprising therefore that in spite of the determined efforts of the government, poaching is still in vogue and except for a few stray cases which are brought to light, the rest go scot free, unbridled by any law of the land.

The data of the 18 years (1975-1992) which are available with the wildlife department of the state and are under review exhibit an incessant practice of poaching. Besides there may be other few cases which have not been recorded and are thus not under review.

Starting from poaching of 7 animals in 1975, the 1992 figures show 9 animals. The peak years of poaching have been 1979 (16), 1981 (29), 1982 (27), 1985 (12) and 1989 (20). The average of all these 18 years shows that the rate of poaching remains at least 10 animals/year which is very high from a small sanctuary covering only 392 km^2 area and having limited species (Fig. 4.3).

Mostly the poachers kill tigers, whose skin, claws, bones and flesh are highly valued in East Asia, and its cooked meat is a delicate recipe, taken for potency. The other animals which fall a prey are Sambhars, antelopes, Chinkaras, deer, bucks etc. which are poached for meat delicacies, specially by the Rajputs, Muslims, tribal people and people of high aristocracy.

Tiger is in red data book. Its speies all over the world are languishly decreasing, partly on account of inbreeding and mostly due to poaching which is a matter of concern. The records of Ranthambhore Sanctuary project that tiger population has almost remained constant, ranging from 35 to 42 throughout the last 25 years, since the Project Tiger was started here. The recent counts of 1996 also give a discouraging figure of 38 tigers inside the Park which is most blemishing and their further increase is still bleak.

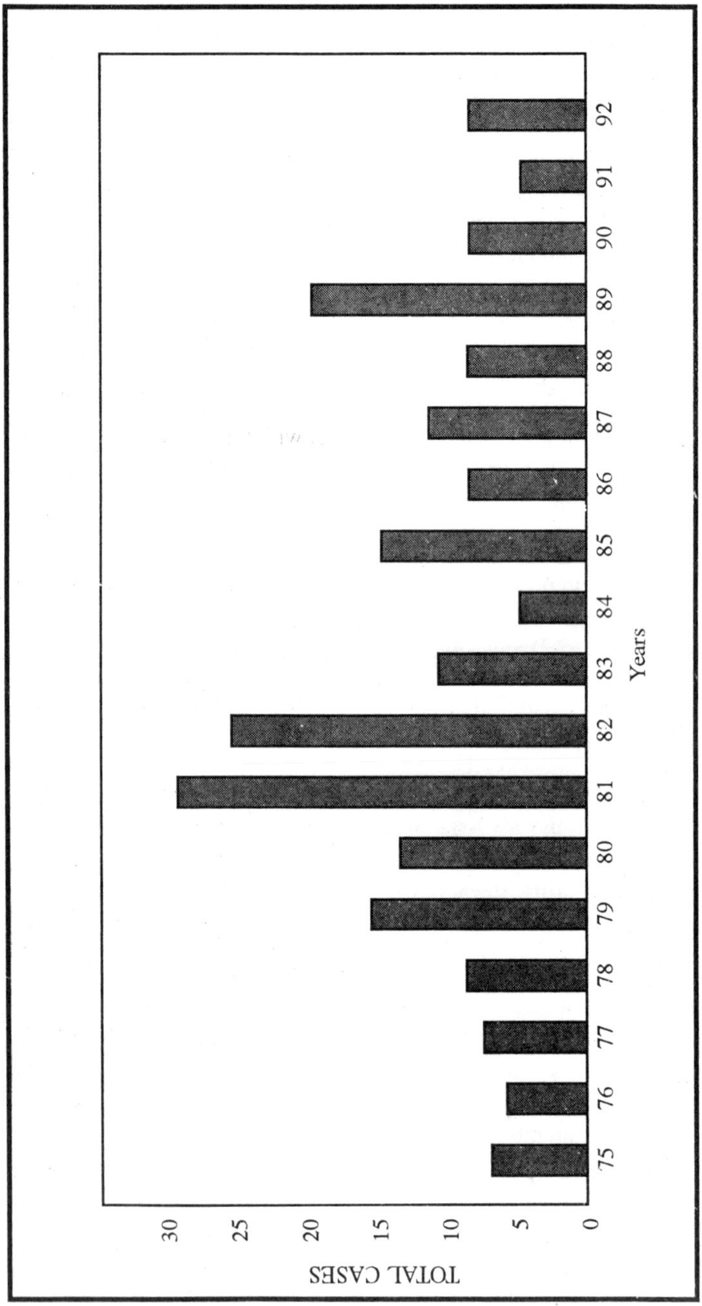

FIG. 4.3 : *Ranthambhore Sanctuary : Poaching and Illicit Killing of Wild Animals (1974-1992)*

It has therefore to be assiduously ascertained that poaching and illicit killings are altogether stopped. The forest guards, although they perform their duties of vigilance most earnestly, yet become mute and docile in front of the high dignitaries, professional poachers and for men in arms.

People's Response and Perceptions to Displacement of Villages

It is significant to evaluate the perceptions, experiences and response of the affected people in regard to the displacement of their original abodes. It is natural that people, living at a certain place for ages, do not want to shift to the new sites, as they are accustomed to their old life-styles, occupations and surroundings. Even if providential exigencies, like floods, droughts and landslips, demand an immediate evacuation of people from their original site of living, the people return to the same place after some time when normalcy returns. It is implicit, therefore, that the people living in the villages inside the Ranthambhore Tiger Sanctuary have evacuated their original abodes after much resistance and to their great dislike.

Experience of Displacement

A sample survey of two newly established villages viz. Gopalpura and Kailashpuri in respect to the experiences of displacement of the displaced people was therefore conducted (Table 4.3). The responses of individual households were noted and generalized for the present study.

It is discernible that the residents of these two villages, who have now come to live here after being shifted from the villages inside the sanctuary, have a mixed response about the change. The residents of both the villages experienced that they had to change their occupation after coming to this new site. Most of them were originally livestock-raisers, supplying milk to the Sawai Madhopur town and had dairy farming as the principal occupation. But now they have become cultivators, as the Government has allotted agriculture land to them.

The residents of both the above villages feel that there has been a decrease in their dependence upon forest resources, which in some cases was substantial while in others only marginal. While inside the sanctuary they had a total dependency upon the fuelwood and animal fodder which was easily available, they have now to spend for these articles.

In regard to their displacement procedure, the people of Gopalpura

TABLE 4.2 : **Sample Villages: Perception about Displacement**

Village Surveyed		Indala	Mordoongri	Bodal	Katholi	Padra
Awareness about displacement	Yes/No	Yes(100%)	Yes(100%)	No(50.74%)	Yes(100%)	Yes(100%)
Perception towards displacement	positive/negative	Negative(75.0%)	Negative(58.7%)	Negative(67.16%)	Positive(52.0%)	Positive(56.4%)
If existence of conflict among the villagers on the displacement issue		No	Yes	No	No	No
Awareness about new proposed site	Yes/No	No	Yes	No	Yes	Yes
Involvement in resettlement process	Yes/No	—	Yes	—	Yes	Yes
Effect on normal life watched out after hearing about displacement	Yes/No	No(68.75%)	Yes(57.1%)	No(100.0%)	Yes(86.4%)	Yes(72.6%)
Future of community if remained here	Profitable	(75.0%)	(63.5%)	(79.1%)	(49%)	(38%)
	Improfitable	(25%)	(35.5%)	(21.9%)	(51.0%)	(62.0%)
Supposed effect	Ties to Land/Forest	feel effected (56.25%)	feel not effected (55.5%)	feel effected (61.19%)	feel effected (68.3%)	feel effected (59.5%)
of Resettlement	Ties to neighbouring villages	feel effected (50%)	feel effected (60.3%)	feel effected (79.10%)	feel effected (52.7%)	feel effected (57.2%)
will be disturbed	Ties to nearby market	feel not effected (68.75%)	feel effected (57.1%)	feel effected (62.68%)	feel not effected (72.4%)	feel not effected (69.89%)
will not be disturbed	Ties to religious locations	feel not effected (62.5%)	feel not effected (74.6%)	feel effected (65.67%)	feel effected (80.7%)	feel effected (76.3%)
Remembrance of previous displacement favourable/unfavourable/Neutral		unfavourable (93.75%)	unfavourable (65.1%)	unfavourable (56.71%) No Response (25.37%)	favourable (55.5%)	favourable (62.1%)

TABLE 4.3 : **Experiences of Displacement (Based on Sample Study)**

Village Name	Gopalpura	Kailashpuri
If noticeable changes are seen in occupational structure — specify	Yes Dairy to cultivator	Yes Dairy to cultivator
Dependence on Forest resources for fodder, fuel etc. — increase/decrease	Decrease	Decrease
Participation in displacement process	Yes	Yes
Nature of displacement Procedure — By force/in friendly manner/As usual/ Timely Scheduled/Sudden Movement	By force Sudden Movement	As usual Timely Scheduled
Level of satisfaction to compen-sation — rich/moderate/poor	Moderate	Rich
View on resettlement issue — it is necessary/could be ignored/ no heavy impacts	Could be ignored	No heavy impacts
Development opportunity at new site more/less	More	More
Remembrance of displacement — cherish occasion/routine type/ worst moment	Worst moments	Routine type

complained that it was sudden and by force; while those of Kailashpuri said that it was time-scheduled and in a usual way. Although the government and park authorities had notified much in advance about this displacement, the villagers inside the sanctuary never gave a placid thought to meet with the requirements of the government which are mandatory. It is for this callous attitude that some people think that it was a sudden outburst and that they had to leave their abodes suddenly.

The compensation offered by the government to vacate the villages has been substantial as all the evacuees of Kailashpuri seem to be totally satisfied while of Gopalpura say that it is moderate. The impact of displacement has not been heavy in all respects, both economically and culturally, because all the evacuees settled in these two new villages are of the same place and with the disbursement of compensation feel economically satisfied.

The respondents feel that in these new sites of settlement there are mroe opportunities for development than before. Truly, it has been a blessing in disguise as most of them feel that they have better avenues for employment in these new villages, which are nearer to the Sawai Madhopur town. Remembrances of the past days are still afresh in their minds and they feel that displacement was the worst moment in their life, when they left the motherland of their forefathers. It is true for any human being but quite likely, they will forget the past and adjust to the present for a better future.

Perceptions about Displacement

The people of villages (Indala, Mordoongri, Bodal, Katholi and Padra), which are yet to be evacuated from the sanctuary, were also interrogate about their perceptions of displacement. Nearly every household in these villages, except those of Bodal (50.7%), knows that they are to vacate these sites for the sanctuary. But the people of 3 villages viz. Indala, Mordoongri and Bodal had a negative response while the remaining two had a positive response. It is usual for these respondents to imagine that after vacating these sites, they may or may not get better sites to live and that their means of livelihood are likely to be greatly affected. Security of employment, their cohesion with new people and the future of their children looms large before their eyes which shows lacks of faith and sincerity in the implementation of government plans. At the same time there is no conflict amongst the villagers themselves about the issue of displacement except in Mordoongri village (Table 4.2).

The households of Indala and Bodal villages showed their ignorance about the new site of their settlement while in the rest three villages, they know in advance about their future site of settlement. It is always good to know about the future places of settlement so that the people can plan for them accordingly. Ignorance always inculcates false and horrified imaginations and so it will be essential to tell these people about their future sites.

Most of the people were found to be involved in the resettlement processes which means they have full knowledge about the place of resettlement and the time to vacate the old dwelling. Also they know about the compensation likely to be disbursed to them and the new vocation they are likely to adopt. A large percentage of these respondents feel that there will be effect on their natural life even after displacement but some of the people in Mordoongri, Katholi and Padra did not accept it. A large percentage of people of these three villages felt that it will bring about a drastic change in their normal life.

It is for this reason that almost in all these 5 villages, a large percentage of the people feel that it will be profitable to live at the original site where they are established for ages. The supposed effects on the life are by and large the ties to their lands, religious sites which are inside the park and close affinities to their neighbouring villages (for details, see Table 3.7). The narrations of previous displacements are unfavourable and this is the main reason that they do not want to leave their old abodes. Reactions in this regard are of mixed nature. The majority of the people of Katholi and Padra tend to have favourable narrations of displacement which took place earlier.

It is amazing to find that a large percentage of the respondents have a positive view about family welfare programmes. Although they have been living inside the boundaries of the sanctuary for long, yet as a result of frequent conversations with the people in and out of the sanctuary, they know about the importance of family planning. A few respondents who did not respond perhaps felt shy of replying to these interrogations.

Environmental awareness is generally found among all the respondents and in all the 5 villages. This is a good omen. Perhaps it is on account of their frequent talks with the forest guards and the authorities that they know about the significance of environmental management, tender care of the wild animals and the need to protect both plants and wildlife.

TABLE 4.4 : **Cattle Population (1988)**

S.No.	Village Name	Cows and Oxes	Buffaloes	Sheep and Goats	Ponies, Horses, Mules and Donkeys	Camels
1	2	3	4	5	6	7
1.	Olwasa (Sawai Madhopur)	528	323	351	48	—
2.	Endwa	784	525	937	28	—
3.	Shyampura	1352	850	1300	48	22
4.	Parli	284	168	466	—	6
5.	Bhadlao	309	246	523	—	9
6.	Kundera	887	479	943	66	18
7.	Rawal	990	566	880	46	15
8.	Uliyana	823	565	1274	9	8
9.	Khandoj	112	59	55	—	3
10.	Sherpur	267	152	319	13	4
11.	Kutalpura Maliyan	93	69	318	2	2
12.	Khawa	300	148	548	—	6
13.	Jamool Khera	300	224	117	12	10
14.	Khilchipur	1552	785	1893	44	14
15.	Ramsinghpura	222	120	96	4	7
16.	Khatipura	324	141	536	118	3
17.	Borif	475	259	643	7	4
18.	Madhosinghpura	128	62	70	—	2

Contd.

TABLE 4.4 : *Contd.*

1	2	3	4	5	6	7
19.	Nimli Khurd	314	412	701	2	79
20.	Nimli Kalan	294	337	731	7	3
21.	Kalibhat	39	38	152	—	3
22.	Kushalipura	88	201	341	—	—
23.	Hajjam Kheri	213	130	375	—	1
24.	Halonda	411	222	546	2	3
25.	Mordoongri	157	282	329	1	—
26.	Bodal	275	271	492	1	6
27.	Bheronpura	91	194	387	—	—
28.	Dumoda	366	292	579	5	7
29.	Baso Kalan (Khandar)	371	176	360	4	1
30.	Baso Khurd	361	217	474	1	—
31.	Bhoori Pahari	985	496	3245	22	25
32.	Dungri	720	257	937	70	7
33.	Sawtan	495	324	1223	6	8
34.	Katholi Dhanapala	154	155	451	6	—
35.	Padra Nizamat	232	263	354	—	—
36.	Goth Bihari	391	100	748	—	—
37.	Sanbar	73	5	58	—	—
38.	Khandar	1359	463	2728	46	8
39.	Rawrah	171	40	237	—	—

Contd.

TABLE 4.4 : *Contd.*

1	2	3	4	5	6	7
40.	Jaisinghpura	299	156	276	—	—
41.	Mei Kalan	618	265	908	—	—
42.	Indala	72	61	279	—	—
43.	Jetpur	168	54	133.	—	6
44.	Chhan	1535	1198	1343	9	18
45.	Allapur	290	201	561	—	7
46.	Bahrawada Khurd	704	477	1259	4	4
47.	Niwandi (S.M.)	186	156	642	—	1
48.	Bhundeda	232	195	802	—	2
49.	Jeenapur	241	202	849	—	13
50.	Himmatpur	62	52	214	5	1
51.	Qila Ranthambhore	8	4	10	—	—
52.	Sakhwas	101	82	335	1	1
53.	Gopalpura	—	—	—	—	—
54.	Kailashpuri (S.M.)	—	—	—	—	—

TABLE 4.4 A : **Animal Population (1988)**

S.No.	Village Name	Cattle Values				
		Cows and Oxes	Buffaloes	Sheep and Goats	Ponies, Horses, Mules and Donkeys	Camels
1	2	3	4	5	6	7
1.	Olwasa (Sawai Madhopur)	528	646	175.5	48	—
2.	Endwa	784	1050	468.5	28	—
3.	Shyampura	1352	1700	650.0	48	176
4.	Parli	284	336	233.0	—	48
5.	Bhadlao	309	492	261.5	—	72
6.	Kundera	887	958	476.5	66	144
7.	Rawal	990	1132	440.0	46	120
8.	Uliyana	823	1130	637.0	9	64
9.	Khandoj	112	118	275	—	24
10.	Sherpur	267	304	1595	13	32
11.	Kutalpura Maliyan	93	138	159	2	16
12.	Khawa	300	296	274	—	48
13.	Jamool Khera	300	448	58.5	12	80
14.	Khilchipur	1552	1570	946.5	44	112
15.	Ramsinghpura	222	240	48.0	4	56
16.	Khatipura	324	282	268.0	11	24

Contd.

TABLE 4.4 : *Contd.*

1	2	3	4	5	6	7
17.	Borif	475	518	321.5	7	32
18.	Madhosinghpura	128	124	35.0	—	16
19.	Nimli Khurd	314	824	350.5	2	632
20.	Nimli Kalan	294	674	365.5	7	24
21.	Kalibhat	39	76	76.0	—	24
22.	Kushalipura	88	402	170.5	—	—
23.	Hajjam Kheri	213	260	187.5	—	8
24.	Halonda	411	444	273.0	2	24
25.	Mordoongri	157	564	164.5	1	—
26.	Bodal	275	542	246	1	48
27.	Bheronpura	91	388	193.5	—	—
28.	Dumoda	366	584	289.5	5	56
29.	Baso Kalan (Khandar)	371	352	180.0	4	8
30.	Baso Khurd	361	434	237.0	1	—
31.	Bhoori Pahari	985	992	1622.5	22	200
32.	Dungri	720	514	468.5	70	56
33.	Sawtan	495	648	611.5	6	64
34.	Katholi Dhanapala	154	310	225.5	6	—
35.	Padra Nizamat	232	526	177.0	—	—
36.	Goth Bihari	391	200	374.0	—	—
37.	Sanbar	73	10	29.0	—	—

Contd.

TABLE 4.4A : *Contd.*

1	2	3	4	5	6	7
38.	Khandar	1359	926	1364.0	46	64
39.	Rawrah	171	80	118.5	—	—
40.	Jaisinghpura	299	312	138.0	—	—
41.	Mei Kalan	618	530	454.0	—	—
42.	Indala	72	122	139.5	—	—
43.	Jetpur	168	108	66.5	—	48
44.	Chhan	1535	2396	671.5	9	144
45.	Allapur	290	402	280.5	—	56
46.	Bahrawada Khurd	704	954	629.5	4	32
47.	Niwandi (S.M.)	186	312	321.0	—	8
48.	Bhundeda	232	390	401.0	—	16
49.	Jeenapur	241	404	424.5	—	104
50.	Himmatpur	62	104	107.0	5	8
51.	Qila Ranthambhore	8	8	5.0	—	—
52.	Sakhwas	101	164	167.5	1	8
53.	Gopalpura	—	—	—	—	—
54.	Kailashpuri (S.M.)	—	—	—	—	—

Cattle Units

Cows & Others — 1.0
Buffaloes — 2.0
Sheep & Goats — 0.5
Camels — 8.0

The survey in respect to eco-destructive activities of the sanctuary and the perceptions of the people of 5 villages in this regard clearly show that they do not have concern about the ecological health of the sanctuary even now. The interviewers found that all these people even now wish to have fuelwood collection, fodder cutting, herb collection, animal grazing and also to do agriculture. The villagers' attitude towards illegal entry to the sanctuary, improper conduct to steal the herbal wealth of the sanctuary and to permit their animals for grazing grasses is still stubborn as it was before and that they think it is their right to use and exploit the resources of the sanctuary. Undoubtedly, these people have continued to live in this area since ages and have been practising all these activities which are now thought to be illegal and their claims have historical legacy. Yet it is essential that these people are made to understand the importance of maintenance of this unique ecosystem which has been surviving for long in these isolated ranges of the Aravallis and maintaining some of the rare species of wild animals which India is proud of.

Status of Biomass

In order to assess the impact on the Sanctuary, biomass status has also been studied on the basis of satellite data. The data supplied by the State Remote Sensing Centre, Jodhpur indicate that the study area of Ranthambhore National Park falls in two IRS LISS II scenes i.e. 29/49 A2 and B2. The earlier analysis for the year 1992 and 1994 was also carried out using satellite data of February month. To have compatibility with previous analysis, the data of nearest possible data are acquired.

The forest mask prepared in the previous analysis has been used for demarcating the forest area. The stratification approach is used for forest and landuse classification. The cultural features like roads, settlements and rail track are digitized from Survey of India toposheets.

The forest area has been categorised under normalized difference vegetation index which is indicative of greenness in the area. The landuse has been classified separately and further these two maps are merged together to get the composite image showing landuse changes as well as forest land categories.

Result and Interpretation

Forest area has been categorised in four classes i.e. good biomass, moderate biomass, less biomass and forest blank. The forest blank area is devoid of vegetation, while good biomass area contains green biomass. Other two categories, i.e. moderate biomass and less biomass, have comparatively less green biomass in that order, which may be due to various reasons.

The satellite observations of greenness depend upon the canopy/foliage of trees/plants.

Comparative study of 1992, 1994 and 1996 images show that the water bodies in the area had less water in 1994 as compared to 1992 and 1996. This surface water condition can be one of the indications of the precipitation/rainfall.

The following products are generated to show the locational changes in forest categories :

A. Landuse/landcover and forest map (Fig. 4.4) for February, 1996.

B. Map showing the changes during 1996 and 1994: It is categorized in six classes, which are self-explanatory. The positive change category indicates that there is an improvement in those areas. The improvement may be from any lower category to higher category e.g. forest blank area (1994) to less biomass (1996) or moderate biomass to good biomass for 1994 to 1996 situation.

C. Map (Fig. 4.4) showing the changes during 1996 and 1992, the explanation in B holds good for this too.

D. Map (Fig. 4.4) showing the changes during 1992-94-96. These show the changes in the three years of observation. The positive and negative changes are shown as follows:

Positive Changes :

a. Positive change from 1992 to 1996 or

b. Positive change from 1994 to 1995

Negative Changes :

a. Negative change from 1992 to 1996 or

b. Negative change from 1994 to 1996.

This means the area which was in moderate biomass category in 1992, while in less biomass category in 1994 will be treated as positive, if the same is classified in moderate biomass category in 1996. Similarly,

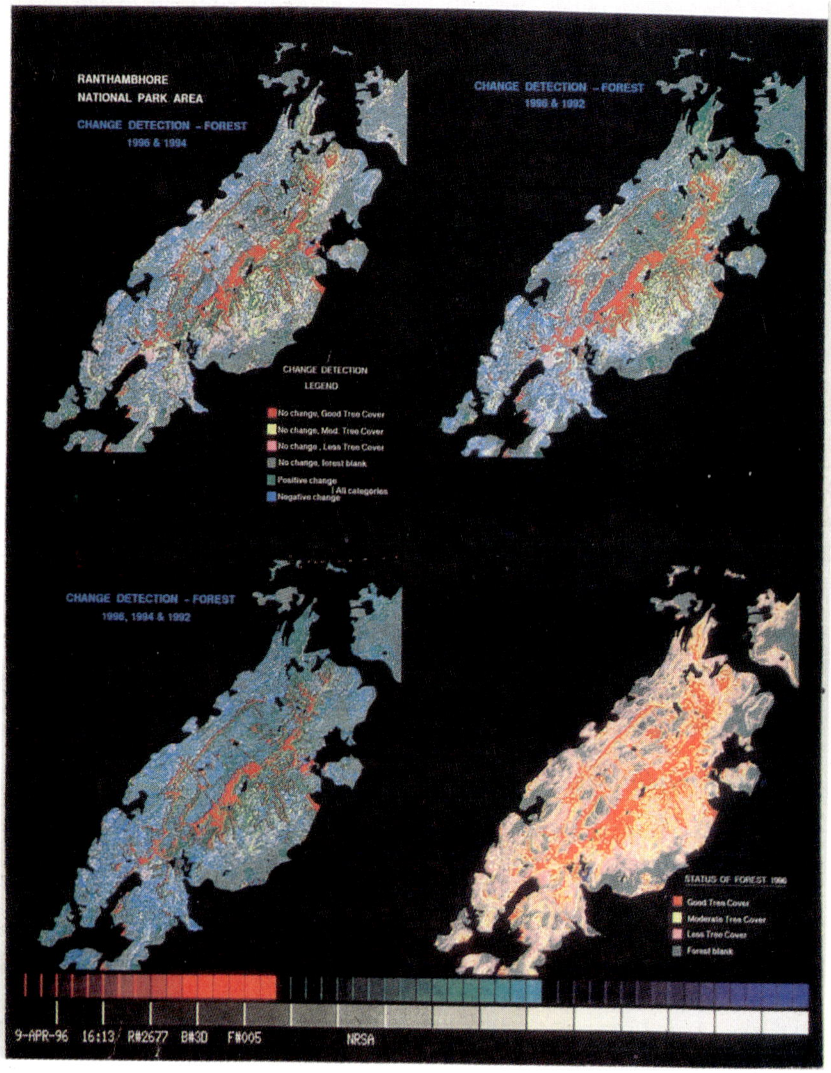

FIG. 4.4 : *Ranthambhore National Park Area — States of Biomass depicting Changes on IRS LISS II Data*

the area which was classified as moderate biomass in 1992, while in good biomass in 1994 will be treated as negative if the same is classified in moderate biomass in 1996.

Some of the observations are given below:

a. Positive change does not necessarily mean increase in treecover but mainly accounts for increase in foliage, tree canopy.

b. Negative changes are observed in fringe areas and major positive changes are in the central area.

c. Spread of water bodies gives the impression that 1996 is better rain fall year in comparison to 1994.

d. Considering the above facts the negative changes are worth noting.

Tables 4.5 A and 4.5 B, 4.6 and 4.7 showing the results for all three points of time are given below.

TABLE 4.5 A : **Forest Area**

Category	Area in hectares		
	1996	1994	1992
Good biomass	13,903.88	12,526.87	13,471.40
Moderate biomass	11,614.36	11,536.00	10,797.23
Less biomass	14,119.14	15,623.28	14,128.34
Forest blank	22,265.67	22,205.27	23,332.40
Others	156.43	163.06	330.11

TABLE 4.5 B : **Changes in Forest Area**

Category	Changes in Area in hectares	
	1996-94	1996-92
Good biomass	+1,377.01	+432.48
Moderate biomass	+78.36	+817.13
Less biomass	–1,504.14	–9.2
Forest blank	+60.4	–1,066.73

TABLE 4.6 : **Landuse/Landcover 1996**

Category	Area in hectares
Crop	45,991.67
Non-cultivated	27,353.77
Scrub/Barren	53,080.40
Surface Water	2,911.46
River sand	2,328.65

TABLE 4.7 : **Changes in Biomass**

Changes in biomass in hectares	1992 to 96	1994 to 96
Good to moderate biomass	2483.01	2513.20
Moderate to less biomass	2679.61	2354.18
Less biomass to forest blank	3862.60	4920.91
Moderate to good biomass	2999.07	3509.57
Less to moderate biomass	3224.45	4063.61
Forest blank to less biomass	5107.54	3922.34

5

CONCLUSIONS AND RECOMMENDATIONS

RANTHAMBHORE National Park/Sanctuary in Sawai Madhopur District of Rajasthan is one of the country's finest tiger reserves. In the scrub forests of this sanctuary it is comparatively easier to sight a tiger than in the Corbett National Park or the other celebrated tiger reserves of the country though the population is just around 38 in a park of 274 square kilometres.

The study of eco-perception of Ranthambhore Sanctuary reveals that this Sanctuary was established primarily to conserve natural and cultural resources for the benefit and enjoyment of tourists and future generations. Fulfilling this dual mission has become complex with time. To a park authorities, sanctuaries and parks are to conserve and preserve the cultural and natural heritage of the nation and on the other hand these protected areas are means of livelihood for the 'gateway communities'. Under such circumstances, where national parks acquire different roles, the park management is faced with the most difficult task of making the management policy for the park, so that everyone's interests are safeguarded.

The study reveals that negative impacts of the people on the Sanctuary are unauthorized grazing by villagers, heavy pilgrim and tourist pressure, dependency of villagers on some of the water bodies on the fringe of the National Park and collection of nonwood forest products (NWFP) and stones lead to disturbance of the ecosystem. On the other hand the people in the surrounding areas are of the opinion that as no major industry, road network, irrigation dams are allowed in the Sanctuary, their living standards have gone down drastically. Crop raiding and cattle lifting by wild animals is another serious problem faced by inhabitants living in an and around the Sanctuary. The restriction of access to forest area inhibits their ability to earn and to meet their basic needs.

In view of these impacts the major management issues for the Ranthambhore Sanctuary are the controlling of grazing and the stopping of fuelwood collection. Although grazing is permitted in the buffer zone, the density is now so high that it degrades the buffer zone and puts heavy pressure on the core area. The data collected during field study on fuel and fodder consumption show that some measures have to be taken immediately to contain this pressure. There is also need to develop alternate sustainable sources of water and Nonwood Forest Products and to regulate tourism.

Grazing

Though animal husbandry is not the major occupation of the people in the majority of the villages, a large number of heads of livestock are kept as supplement to agriculture. Village grazing lands have partially or totally been converted to agriculture and the remaining ones do not have the potential to sustainably provide fodder and grasses to the large population of cattle. In this situation, the Ranthambhore Tiger Reserve forest comes under increasing grazing pressure, with livestock remaining in the forest from the onset of the monsoon. The possibility of disease transmission to wildlife remains very high, as the cattle are not inoculated against common diseases. The competition between wild animals and livestock increases and often, the wild animals have to confine themselves to a smaller area. The intense pressure from livestock reduces the regeneration potential of the forest. This can be found out by low density of vegetation in certain areas which have been heavily exploited. This is further aggravated by the appearance of weeds in some areas, which replace the palatable species.

Apart from the local livestock population, large itinerant herds also make use of the forests. Sheep and goat graziers from far-off places encroach upon the forest from March to June. These animals carry diseases which may be transmitted to the wildlife. Competition for food also increases in the lean period. The buffer area of Ranthambhore Tiger Reserve, especially Kailadevi Sanctuary, has heavy pressure of grazing from migratory sheep. Approximately 1.5 lakh sheep come to the area, in two seasons. As the Sanctuary is adjacent to the protected forest area where these graziers have grazing rights, ingress of sheep in the Sanctuary is a constant nuisance. Even the local villagers detest these sheep but can do nothing against the organized Raibaris, who are migratory sheepherders.

The village pastures (revenue lands) which once provided grass support are severely degraded or illegally occupied in a majority of the villages. Keeping large herds of goats and sheep is a recent development around the Ranthambhore Sanctuary, and it is because of this that the demand for fodder has considerably increased. Green fodder is collected from July to September, and dry fodder from December to January. *Anogeissus pendula* forms the major fodder species, along with *Butea monosperma, Ficus tomentosa* etc. *Apluda mutica, Aristida spp., Bothriochloa pertusa* etc. form the major grasses collected by the people. People in several villages and especially in the surrounding townships sell the fodder and grasses collected from the forest, during summer, when the avilability of fodder depletes.

Eco-Development Strategies

(a) Animal Health Care Programme aimed at reducing numbers and improving breeds to facilitate stall feeding.

(b) Fodder plantations in private lands and alternating closures of common lands to help regeneration for fodder harvest.

(c) Agro-Forestry.

Collection of Firewood

Seasonal unemployment is to a great extent the reason for increased dependency on the forest. When villagers do not have any other means of getting income during the period when their lands remain fallow, they begin to depend on the nearby forests for sustenance. Interviews of villagers have revealed that if proper employment facilities are provided to them, they would desist from exploiting the forests for fuelwood. Firewood is the main source of fuel in small towns surrounding the Ranthambhore Sanctuary, such as Sawai Madhopur, Man town (a suburb of Sawai Madhopur town) and Khandar. A large number of inhabitants of these three towns are also involved in selling firewood. The reason for this high dependency on firewood is because only 1% to 2% of the population has gas connections and animal dung is usually put to decomposing for use as manure. Also, kerosene oil is always available, especially at controlled prices.

The average farmer's landholding is small and the production of crop residue is not enough to meet the local fuel needs. Also, crop residues are available for only a short time in a season. All these factors contribute to the extensive demand for firewood in the rural areas. Mostly women collect firewood. The present data indicate that in villages mass collection of firewood, for their own consumption, takes place in winter after Dipawali (November). The firewood collected during this time is stored for the whole year. On an average, 5-8 kg of wood is consumed per day in a household with 7-8 members, in addition to some dung cakes and crop residues. Those involved in the sale of firewood, collect it on a daily basis throughout the year. Usually dried, fallen wood is collected, but many a time forest wood is extracted too. Most species can be used for this purpose, but *Anogeissus pendula* is the preferred firewood. Other species, like *Butea monosperma, Maytenus emarginatus and Acacia leucophloea,* are also used, but in small quantities. A severe shortage of dried fallen wood in the peripheral areas has compelled people to encroach further on the forest. The distance covered by a woman to collect firewood from the forest has increased manifold over the years. This reflects the impact of firewood collection on the forest, and the impact of the shortage of firewood on the people, especially the women.

Eco-Development Strategy

(a) Several of the activities suggested to address this need are closely related to what can be considered as agro-forestry. They are fuelwood and agro-silvicultural plantations, horticultural plantations with inter-cropping, bundh plantations around private land, soil conserving measure where grasses, bushes, or trees are incorporated in the vegetative bunding of soil.

(b) Introduction of fuel-efficient stoves.

(c) Biogas installation: The KVIC model is quite successful in this area (CEE report). To propagate the idea, demonstrations will have to be done by Sanctuary management.

(d) Income-generation activities.

Timber

It also forms one of the major requirements of the people from the forest.

It is needed for making agricultural implements like ploughs, for house construction, for fencing on the fields, and for constructing storehouses for dry fodder, raised platforms in the fields to guard the crops from wild animals and other items of daily use like chairs, tables and cots. *Anogeissus pendula* forms the major timber species. Other species are *Butea monosperma, Acacia leucophloea* and *Acacia nilotica.* Timber is mainly collected by the menfolk. Timber extraction is not allowed in the Sanctuary but nevertheless it is done illegally.

Water Resources

Lowering of the water-table and scarcity of water for cattle as well as human beings has put a tremendous stress on the water bodies with the Sanctuary. The use of water sources within the study area, by the livestock, increases the chances of spread of epidemics among wild animals, and disturbs the wild animals' habitat. The runoff which drains out through seasonal rivulets can easily be stored and utilized for irrigation purpose during lean period. This can be done by repairing existing tanks, increasing their capacity by removing the silt and construction of new tanks.

Eco-Development Strategies

(a) Soil and water conservation works like checkdams and afforestation of degraded mini watersheds inside the Sanctuary will help in raising the water-table.

(b) Lift irrigation schemes of Olwara, Piplait and Pipalda on two river systems of Banas and Chambal are the most potential alternatives to produce enormous biomass outside the Sanctuary to cater to the needs of surrounding population.

(c) Digging of wells along the major lineaments.

Crop Depredation

Crop damage and crop raiding has been projected as one of biggest problems of the villages adjoining the Ranthambhore Sanctuary, as an estimated 30% of the produce is eaten away by wild animals. Crop depredation by wild animals also creates human-animal conflicts, and sometimes results in the villagers killing the animals. There are also incidents of cattle lifting by tigers and leopards.

Eco-Development Strategies

(a) Electric and green fencing and stone walls for crop protection. Village eco-development committees should take up the responsibility of maintaining the fence.

(b) Increasing involvement of local people, especially the Moghyar tribals, in patrolling and crop protection.

(c) Crop Insurance.

(d) Development of innovative methods, including biological methods, for crop protection.

(e) Payment of compensation for crop damage and livestock injury and death.

Tourism

The enormous increase in tourist traffic is a significant disturbance to animals and to the habitat. Tourism is presently confined to the core zone. There is a need to identify suitable, alternate areas to divert the pressure. It is also necessary to organise tourism in a manner that helps the local people to earn a living rather than making profits for corporate houses and rich entrepreneurs. The tourism activities must be ecologically sensitive and the thrust must be on eco-tourism.

The study indicates that the tourist traffic is likely to increase in future and so also the vehicular traffic. While it may not be possible to shut the doors of the Sanctuary for the welfare of the wildlife it should be possible that electric vans which are noiseless and pollution-free, cable-cars in between the mountain ranges and balloon flights may be arranged for the visitors as an alternative to diesel-driven vans. It would be more safe to the visitors also to save themselves from any possible attack of wild animals which at times are in privacy and siesta.

Eco-Development Strategies

(a) Ranthambhore Sanctuary has great potential to offer employment to the local people through tourism. In this direction efforts have already started. A nature guide training camp was organised and 22 youths from the neighbouring area of the sanctuary have been trained as nature guides. These guides accompany the tourists and can earn Rs. 50 per day.

(b) Villages near the tourism zone and even on the periphery could easily be developed as handicraft centres, catering to the tourists coming to Ranthambhore National Park. Sales outlets can be provided at the booking office or interpretation centre.

(c) Kailadevi Sanctuary has vast tourism potential, which can be harnessed if a tourism area is developed by repairing roads and patrolling paths.

Pilgrimages

The pilgrim traffic has also caused problems in the past and continues to be a disturbance. This is a sensitive issue and needs to be facilitated through the cooperation of the pilgrims. This can be solved by organising awareness programmes for the pilgrims, stressing the value of the forest and tiger and the way to protect them.

Poaching

This has become a very sensitive issue in this park. There is reason to believe that some of the hunting tribals have been lured to poach for economic gains. Policing such a vast area, which is surrounded by dense population, becomes very difficult.

Strategies besides constant vigil that should be resorted to are :

(a) Better wireless communications and more vehicles are needed for patrolling.

(b) Intensive patrolling by anti-poaching squads needs to be undertaken. Such squads should involve the local people. In fact, there should be a much greater involvement of the local people in activities related to the prevention of poaching.

(c) A 'reward scheme' for informers needs to be introduced.

Eco-Development Programmes

Human Resources Development

Human resources development should include various training inputs needed for carrying out the proposed management and eco-development activities. Training for income generation activities should be carried out. Local artisans who are capable of making and selling stoves should be given training, on the principles of fuel efficiency, so that they could use their innovative skill for making locally suited and acceptable stoves and sell them. Training for setting up biogas plants should be given.

Environmental Awareness

Awareness programmes for both visitors and local people focussed on the value of the Sanctuary and its linkages with the regional, national and global conservation efforts, should be started. An attempt to establish an exchange of knowledge between local persons and scientists, where the former should share their understanding of local resources, medicinal and other social significance of plants and animals, the local nomenclature and understanding. The scientists should, on the other hand, share their understanding of the surrounding areas to interpretation centres, screening films, inviting the village elders to share their knowledge and arranging day visits to the Sanctuary for local people so that they also have an opportunity to enjoy and appreciate the Sanctuary. NGOs should be involved for carrying on a campaign to use crop residue better and to improve the livestock condition.

Institutional Structure

It is suggested to have a cooperative or a society which should have, as its members, all the people who are dependent upon the resources of the Sanctuary. This cooperative should organise suitable tourist-related activities and support services.

It is also needed to have village level eco-development committees which should interact with the Ranthambhore National Park authorities and other concerned agencies on implementing and planning site specific eco-development packages.

It is also needed to identify the local NGOs which have been working on rural development issues in the area, so that they can be involved in the planning and implementation of the various biomass and income generation schemes proposed to be undertaken under eco-development.

It is suggested to have an institutional mechanism for enabling the local people to be consulted, and to participate in Sanctuary management.

It is also proposed to have an eco-development coordination committee at the state level, chaired by the Chief Secretary, or in his absence, by the Forest Secretary. This should be supported by a district level coordination committee, convened by the Field Director and involving the local collector chaired by an eminent environmentalist of the state. Such committees should coordinate the inputs of the various line departments like agriculture, animal husbandry etc., for implementing income generation schemes as part of eco-development. The state and district level committees should also help in better coordinating protection measures. The forest authorities are working closely with the two NGOs, WWF, India and the Ranthambhore Foundation. Both organisations work with the villagers. The Global Environmental Facilities and the World Bank will also be pumping in close to Rs. 380 million over the next four years for development of 112 villages around the National Park. The villages within the core area (Indala, Mordoongri, Katholi and Padra) should now be shifted to new site on sound ecological footings with proper compensations.

Ranthambhore has recently been much in the news, nationally and internationally as there is evidence of large-scale tiger poaching. The proposed strategy for eco-development and management inputs will go a long way in making the Tiger Reserve secure.

It would be appropriate here to refer the work of Ranthambhore Foundation. The Ranthambhore Participatory Rural Appraisal (PRA) was done by the Ranthambhore Foundation team headed by Nitin Pandya. They covered 11 villages in all, namely, Halonda, Kuthalpura, Olwara, Talwada, Kushalpur, Gopalpura, Jethpur, Mordoongri, Kailashpuri and Bassukalan. The PRA techniques used were (a) timeline, (b) social map, (c) resource map, (d) seasonality, (e) services and opportunities, (f) preference ranking of problems and (g) preference ranking of solutions and their modes. The following issues emerged through the PRA exercise:

A. People's dependency on the Sanctuary was on the following resources given in descending order of priority:

1. Firewood

2. Grazing Grounds

3. Timber

4. Grasses and Fodder

5. NWFP, Butea leaves, Limestones, Tendu fruits and leaves, sand,

rocks

6. Land

B. The impact of the sanctuary on the people was in the following ways :

1. Crop damage

2. Cattle lifting

According to the PRA, the village communities are willing to unite, through village committee, to solve some of their basic problems. The villagers also felt that greater income generation opportunities for the women was a priority.

Approach Suggested

- To work with clusters of villages and to tackle problems at village level through village committees.
- To tackle first those issues which are considered priorities by the local people. This would help in getting the trust of the villagers.
- To orient the forest staff, and to motivate them, to involve the people in management and protection of the sanctuary.
- To organise trips for villagers to surrounding areas (NGO/Govt. Projects) for exposure to biogas, breed improvement, soil/water conservation work, pasture land improvement and viable income-generating schemes.

Solutions and Strategies Recommended

- Any eco-development measures must be carried out only after PRA has been conducted in the particular village.
- Arresting surface runoff to recharge water-table. This can be done by construction of 'checkdams', anicuts, contour bunding on the basis of the PRA work done. This can also be an initial rapport-building exercise.
- Establish income generation activities for people who are dependent on the forest.

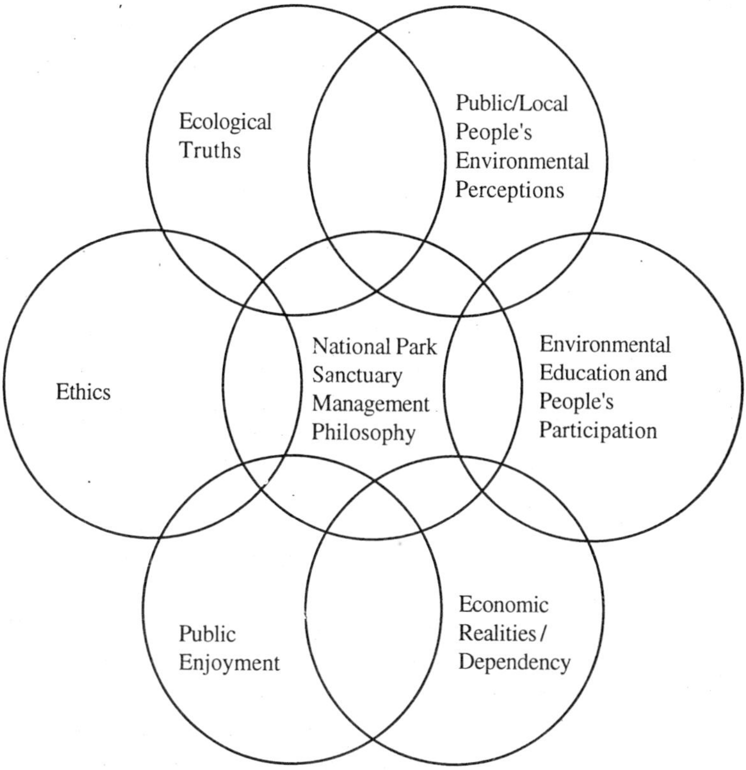

(Modified from Keiter, 1991)

FIG. 5.1 : *Spheres of Influence on Park Management Philosophy*

Plate 9. Pace towards Sanctuary.
Every morning camels from outside are being entered in Sancuary for whole day.

Plate 10. A camel herd, near Moordoongri.

Plate 11. Beauty of simplicity. All utensils are being hung as showpieces, Moordoongri.

Plate 12. A very identical rural set-up, Moordoongri.

Plate 13. A farmer with his traditional friend (plough)
All agriculture of Indala, depends on only rain water in absence of any arrigation source.

Plate 14. Kachacha Talav, Indala.
No use for irrigation because of longer distance from cultivated land, absence of electricity supply and lack of water storage measures.

Plate 15. An open conversation with the people of Indala.

Plate 16. We have our own distinct life.
People are facing hundreds of kinds of problems, here at Indla.

Plate 17. Evidence of history.
Ruin of old Indala village, near Indala.

Plate 18. Lok Devtas' denial to go anywhere, so they are still worshiped at the old site after several decades. Old sitee, Indala.

Plate 19. Hanuman Temple, at old site, Indala.

Plate 20. Cattle grazing on forest land is very common scene everywhere, near Indala.

Plate 21. We will pull out the water from hearts of stones.
A well digging practice at an altitude about 200 mts, Indala.

Plate 22. A well-to-do something new.

A farmer tried once with a diesel pumpset to irrigate his fields by Kachcha Talav near village. But he is not satisfied with its results.

Plate 23. Bawdi, behind the Hanuman temple, old site, Indala.

Plate 24. The Tanka, the only source of potable water, at Indala.

Plate 33. Main road, crossing village, some tea shops and general stores can be notices, Bodal.

Plate 34. A diesel pumpset may have many uses in rural areas. A flour mill, a water tap, a fodder cutter all are being operated by the same set.

Plate 35. Miserdarra Booking centre, Ranthambhore National Park.

Plate 36. Quartzite scarp carved out by Nature, Ranthambhore National Park.

Plate 37. Kachcha Houses, Gopalpura, (New shifted establishment).

Plate 38. An Eco-method of geting fuelwood, generally used, it cannot destroy trees but helps in their healthy growth.

Plate 39. Village at distance, undeveloped surroundingsshow infantile stage of the new new village Gopalpura

Plate 40. Man showing his allotted field in compensation — undulating poor waste-land

Plate 41. Talai, constructed by Panchayat with cooperation of local people and W.W.F., Gopalpura

Plate 42. Afforestation to fulfil local needs by W.W.F. in association with forest department, Gopalpura

Plate 43. Government Primary School, Gopalpura

Plate 44. Brick kilnnear Gopalpura

Plate 45. Ranthambhore Sevika
A medical centre established by Ranthambhore Foundation, at the west outskirt of National Park

Plate 46. Mud wall to stop and collect rain water for irrigation purpose, Cart truck to Katholi.

Plate 47. Where is forest?
Who could believe that these far extended fields are also part of the National Park, from cart track to Katholi.

Plate 48. Arhar (purse) cultivation along the cart track to Katholi.

Plate 49. Are we waiting for our turn?
Near Katholi and Padra.

Plate 50. Where they have gone?
Deforestation near Katholi and Padra

Plate 51. Dev Narayan Temple, located on the hill side, Katholi

Plate 52. Dev Narayanji, the lok Devta commony worshipped by Gurjars

BIBLIOGRAPHY

Azeez, A.P., 1992. "Conservation and Socio-economic Conflicts". *The Green Age,* Vol. 1, No. 1, June, pp. 26-29.

Bagla, P., 1992. Closer to Conservation. *Sanctuary Asia,* Vol. XII, No. 4, Bombay.

Centre for Science and Environment, 1982. *The State of India's Environment : A Citizen's Report.* Centre for Science and Environment: New Delhi.

Centre for Science and Environment, 1985. *The State of India's Environment: The Second Citizen's Report.* Centre for Science and Environment: New Delhi.

Cherail, K., 1993. "Wildlife Conservation Strategy: Time to Change", *Down to Earth,* November 30, New Delhi.

Dixon, J.A. and Sherman, P.B., 1990. *Economics of Protected Areas: A New Look at Benefits and Costs.* Earthscan Publication Ltd., London.

Ferguson, Gary, 1995. *Walking Down the Wild,* Karper Collins West, New York, p.170.

Forestry Survey of India, 1987. *The States' Forest Report.* Ministry of Environment and Forests, Government of India, Dehradun.

Gee, E.P., 1962. "National Parks in India", in Dr. H. Cahalane (ed.), *National Parks : A World Need.* Special Publication M-14, American Committee for International Wildlife Protection, New York.

Ghimire, K.B., 1991. *Parks and People : Livelihood Issues in National Parks Management in Thailand and Madagascar.* United Nations Institute for Social Development, Geneva.

Iyer, Gopal K., 1996. Sustainable Development — Ecological and Social Dimensions, Vikas Publishing House Pvt. Ltd., New Delhi.

IUCH, 1985. *United Nations List of National Parks and Protected Areas,* IUCH, Gland.

Jena, N.R., 1993. "Issues in Parks and Sanctuaries: Some Observations", in PRIA's (1993) *Report on National Workshop on Declining Access to and Control over Natural Resources in National Parks and Sanctuaries,* New Delhi.

Jena, N.R., 1993. "A Critical Analysis of the Legal Provisions of Sanctuary and National Park in India", in PRIA's (1993), *Report on National Workshop on Declining Access to and Control over Natural Resources in National Parks and Sanctuaries,* New Delhi.

Jena, N.R., 1996. National Parks and Sanctuaries vs People's Rights — Some issues of concern in Sustainable Development (Ed.) by Iyer, K.G., pp. 276-291. Vikas Publishing House Pvt. Ltd., New Delhi.

Kannan, K.P., 1983. "Forestry Legislation in India: Their Evolution in the Light of the Forest Bill, 1980", in Walter Fernandes and Sharad Kulkarni's (eds.), *Towards a New Forest Policy: People's Right and Environmental Needs.* Indian Social Institute, New Delhi.

Kothari, A., P. Pande, S. Singh, and D. Variava, 1989. *Management of National Parks and Sanctuaries in India: A Status Report.* Indian Institute of Public Administration, New Delhi.

Kulkarni, S., 1983. "The Forest Policy and Forest Bill: A Critique and Suggestions for Change", in Walter Fernandes and Sharad Kulkarni's (eds.), *Towards a New Forest Policy: People's Right and Environmental Needs.* Indian Social Institute, New Delhi.

MacKinnon, J. and MacKinnon, K., 1986. *Managing Protected Areas in the Tropics,* IUCN, Gland. Ministry of Environment & Forests, 1988. *The National Forest Policy,* Government of India: New Delhi.

Ministry of Environment and Forests, 1992. *Standard Note for Supplementaries for Questions Related to Wildlife Sanctuaries and National Parks: Annexures I and II,* New Delhi.

Natraj Publishers, 1992. *The Wildlife (Protection) Act, 1972.* Natraj Publishers and Traffic India, Dehradun.

Panwar, H., 1993. "Status of Management of Protected Areas in India", Paper Presented at FAO Regional Expert Consultation on Management of Protected Areas in the Asia-Pacific Region, 10-14 December : Bangkok (mimeo)

PRIA, 1993. *Report on National Workshop on Declining Access to and Control over Natural Resources in National Parks and Sanctuaries.* Society for Participatory Research in Asia : New Delhi.

Ramachandran, H., Saxena N.C. and Ramachandran, N., 1994. "Resource Sharing in and around National Parks and Sanctuaries", Discussion paper-1 for Workshop on Issues in Resource use and Institutional Structures in and around International Parks and Sanctuaries, 6-8 January, Mussoorie.

Rao, K., 1988. "Management Planning for Protected Areas in Developing Countries : Project Report". Wildlife Institute of India, Dehradun (Unpublished).

Reed, D. (ed.), 1993. *The Global Environmental Facility: Sharing Responsibility for Biosphere,* Vol. II. Worldwide Fund for Nature International, Washington D.C.

Reid, W. and Miller K., 1989. *Keeping Options Alive: The Scientific Basis for Conserving Biodiversity.* World Resources Institute, Washington.

Reti, I. Osefatn, 1986. "Resolving Conflicts between Traditional Practices and Park Management", PARKS, Vol. 11, No. 1.

Runte, A., 1979. *National Parks : The American Experience.* University of Nebraska Press, Lincoln.

Singh, S. and Rodgers, W.A., 1990. "India : National Parks and Natural Reserves", in Craig W. Allen's (ed.), *International Handbook of National Parks and Nature Resources,* Greenwood Press, New York.

Stephen and Thora Amend, 1992. "Human Occupation in the National Parks of South America: A Fundamental Problem", PARKS, Vol. 3, No. 1, January.

The Times of India, 1993. "Fall in Tiger Count Acknowledged", 27 December, New Delhi.

UNEP, 1989. *Integrated Management of Resources in Africa,* UNEP, Nairobi.

Wells, M.N. and Kartina Brandon F., 1993. *Biodiversity Conservation and Local People's Development Aspirations: New Priorities for the 1990s* (mimeo).

World Resources Institute, 1990. *World Resources 1990-1991.* Oxford University Press, New York and Oxford.

World Resources Institute, 1992. *World Resources 1992-93.* Oxford University Press, New York and Oxford.

Reports

Berkmuller, K.P., Mukherjee, S.K., Mishra, B.K., Grazing and Cutting Pressure at Ranthambhore National Park.

Biodiversity Conservation Through Eco-development - An Indicative Plan for Ranthambhore Tiger Reserve, Rajasthan UNDP, IIPA, New Delhi, October 1994, pp. 151-164.

Champion H.G. and Seth, S.K., Forest Types of India (1968).

Government of Ranthambhore Tiger Reserve (Eco-Development Plan, Drift). Rajasthan, Department of Forest.

Questionnaire filled by Park Authorities.

Rathore, S.S., Management Plan for Ranthambhore Tiger Reserve (1984-89).

Rodgers, W.A. & Panwar, H.S., Planning a Wildlife Protection Area Network in India, Wildlife Institute of India, Dehradun (1988).

Valmiki Thaper (1987-88): A Review of Ranthambhore National Park.

INDEX